Paleontology

In the wake of the paleobiological revolution of the 1970s and 1980s, paleontologists continue to investigate far-reaching questions about how evolution works. Many of those questions have a philosophical dimension. How is macroevolution related to evolutionary changes within populations? Is evolutionary history contingent? How much can we know about the causes of evolutionary trends? How do paleontologists read the patterns in the fossil record to learn about the underlying evolutionary processes? Derek Turner explores these and other questions, introducing the reader to exciting recent work in the philosophy of paleontology and to theoretical issues including punctuated equilibria and species selection. He also critically examines some of the major accomplishments and arguments of paleontologists of the past forty years.

DEREK TURNER is an Associate Professor of Philosophy at Connecticut College. His previous publications include *Making Prehistory: Historical Science and the Scientific Realism Debate* (Cambridge, 2007), as well as a number of articles on the philosophy of paleontology.

Paleontology

A Philosophical Introduction

DEREK TURNER

Connecticut College

CAMBRIDGE
UNIVERSITY PRESS

University Printing House, Cambridge CB2 8BS, United Kingdom

Cambridge University Press is part of the University of Cambridge.

It furthers the University's mission by disseminating knowledge in the pursuit of
education, learning and research at the highest international levels of excellence.

www.cambridge.org
Information on this title: www.cambridge.org/9780521116374

First published 2011

A catalogue record for this publication is available from the British Library

Library of Congress Cataloguing in Publication data
Turner, Derek D. (Derek Donald), 1974–
Paleontology : a philosophical introduction / Derek D. Turner.
    p.   cm. – (Cambridge introductions to philosophy and biology)
Includes bibliographical references and index.
ISBN 978-0-521-11637-4 (hardback) – ISBN 978-0-521-13332-6 (paperback)
1. Paleontology – Philosophy.   I. Title.   II. Series.
QE721.T88   2011
560 – dc22   2010048061

ISBN 978-0-521-11637-4 Hardback

This book is dedicated to my teacher, John F. Post, with gratitude.

Contents

Figures

Acknowledgments

In the spring of 2008, I spent a semester at the University of Pittsburgh's Center for Philosophy of Science, where I began work on this project. I am grateful to the Center faculty, fellows, and graduate students for the opportunity to try out some of the ideas that later made it into this book. I benefited immensely from conversations with Michael Baumgartner, Delphine Chapuis-Schmitz, Richard Dawid, Isabelle Drouet, Mehmet Elgin, Yoichi Ishida, Sandra Mitchell, John Norton, Ed Slowik, and Jim Woodward. My special thanks go to John Norton for making the Center such a collegial and challenging place to do philosophical work.

Over the summer of 2009, I received invaluable research assistance, as well as feedback on drafts of chapters, from Andrew Margenot. Andrew's work with me on this project was paid for by a ConnSSHARP grant for summer research.

This book owes a great deal to comments I received from audiences at the University of New Hampshire, the University of Alabama in Huntsville, the University of Pittsburgh, Florida State University, and at the summer 2009 meeting of the Society for Philosophy of Science in Practice.

A running conversation with Carol Cleland about the nature of historical science has helped to shape my thinking about paleontology. I'm grateful to Rob Inkpen for stimulating conversation and correspondence about historical contingency. Patrick Forber pointed out a mistake I had made in some of my work on trends, and thus saved me the embarrassment of repeating it here. Nick Jones showed me how stochastic processes can have inevitable outcomes. Dan McShea offered helpful critical comments on an earlier paper that contained some arguments that made it into Chapter 8. I also thank Kim Sterelny for his detailed comments on an earlier paper on contingency. Along the way, I have learned a lot of what I know about the paleobiological revolution from conversations with David Sepkoski, as well as from following David's work. My beloved colleagues at Connecticut College – Simon Feldman, Nina Martin,

Andy Pessin, Kristin Pfefferkorn, Larry Vogel, and Mel Woody – have read and offered thoughtful comments on more of my work on paleontology than they probably ever wanted to. These are just a few of the many debts I owe.

I'm grateful to Denis Walsh and Michael Ruse for their interest in this project, as well as for their helpful advice and suggestions along the way. Joanna Garbutt, at Cambridge University Press, and Viv Church, who copy-edited the book, have been wonderful to work with during the production process.

A small portion of the material in Chapters 7 and 8 has appeared in previously published articles. Some of the ideas from those papers are presented here in heavily rewritten form. I reuse this material with kind permission from Springer Science and Business Media: *Biology and Philosophy*, "How much can we know about the causes of evolutionary trends?" 24(2009): 341–357; *Biology and Philosophy*, "Gould's Replay Revisited," published online first on August 12, 2010.

Finally, I thank Michelle Turner for putting up with me while I worked on this, for listening to me go on and on about evolutionary trends and PE at the dinner table, for lots of good advice about writing, and for her impeccable editorial comments.

1 Introduction

Paleontology and evolutionary theory

A world without fossils

Imagine a planet almost exactly like ours, but with one crucial difference: it has no fossils. Call this imaginary planet *Afossilia*. Afossilia and Earth harbor the very same kinds of living things, from ferns to human beings to *E. coli* bacteria. Both planets have the same surface features and the same types of rocks. And both have experienced exactly the same evolutionary histories, with the same species evolving and going extinct at exactly the same time. We can even suppose that you and I have counterparts living on Afossilia – that is, that there are people there who are exactly (or almost exactly) like us.

Some Biblical literalists hold that God placed fossils in the rocks in order to test our faith in scripture. I invite you to join me now in thinking about a simple inversion of this familiar idea: what if God – or if not God, then some more sinister spirit – systematically removed all the fossils from the rocks just before (Afossilian) humans evolved and began to study the world around them. Afossilia has no fossilized footprints, leaf imprints, shells, pollen, teeth, bones, coprolites (fossilized feces), or any of the remains of ancient organisms that we on Earth can see on display in natural history museums.

Suppose you had the opportunity to tour a major research university on Afossilia. There you would find physicists, cosmologists, astronomers, chemists, biochemists, and molecular biologists doing exactly the same things that scientists in those fields do here on Earth. But you would find no paleontologists on Afossilia – no departments of paleontology or professional associations for paleontologists. A world without fossils must also be a world without paleontology ("the study of ancient beings"). The natural history museums, if there were any at all, would contain exhibit halls full of rock and mineral samples, as well as stuffed and pickled specimens of creatures living today, and not much else. *Jurassic Park* never appeared in Afossilian theaters, and Afossilian children have no dinosaur toys or books.

At the end of the day, this thought experiment might not turn out to be fully coherent – a common problem with philosophical thought experiments. Thought experiments often ask us to imagine scenarios that seem logically possible, but which gradually stop making sense if you ask the right awkward questions. You may already be thinking of problems with this one. For example, do the Afossilians have any fossil fuels? What exactly does "fossil" mean in the first place? (I return to that issue at the end of the book, in Chapter 10.) For now, we need not worry too much about the details. An imperfect thought experiment can still serve as a useful device for generating philosophical questions.

What would Afossilian biology look like? And in particular, what would the Afossilians think (or be justified in thinking) about evolution? Would their views about evolution differ from the views of scientists on Earth? Important parts of Afossilian biology would be just like biology here on Earth. For example, the Afossilians would probably have much the same understanding of genetics and microevolutionary change that we do. Nothing would keep them from studying the ways in which natural selection, drift, mutation, and migration can lead to changes in gene frequencies in populations. But what could the Afossilians know about large-scale evolutionary change? Can scientists learn anything about evolution from the fossil record that they simply cannot learn in any other way?

Organismic *vs.* evolutionary paleontology

Paleontologists use fossils to try to understand the history of life on Earth. The term "paleontology" literally means the study of ancient being, and it is sometimes contrasted with "neontology," which refers to the study of life as it exists on Earth today. We can draw a rough distinction between two kinds of research that paleontologists do. The first kind of work, which I will call *organismic paleontology*, attempts to answer questions about the behavior, the biology, and/or the ecological role of some specific type of prehistoric creature. *Evolutionary paleontology*, by contrast, focuses less on questions about specific kinds of prehistoric life, and more on questions about the nature of large-scale evolutionary patterns and processes. This book will deal mainly with evolutionary paleontology.

Figure 1.1 *Tylosaurus dyspelor*. This restoration of a mosasaur appeared in an early paper by Henry Fairfield Osborn (1899), of the American Museum of Natural History.

In order to make the distinction between organismic and evolutionary paleontology clear, it will help to consider examples of each of these two kinds of paleontological research. From there, I will go on to develop some of the questions that will take up the rest of the book.

First, an example of organismic paleontology: the term "mosasaur" means "reptile of the Meuse," and the first mosasaur remains were found in a quarry near the Meuse River in Holland in the late 1700s. The mosasaurs were not dinosaurs, although they flourished during the Cretaceous period, the heyday of the dinosaurs. Instead, the mosasaurs were intimidating marine reptiles – the archetypal prehistoric sea monsters (see Figure 1.1). The very largest ones grew to lengths of 45–50 feet, though most were smaller. The mosasaurs became extinct at the same time as did the non-avian dinosaurs, around 65 million years ago. Their remains have been found in Mesozoic rocks all over the world (Bell 1997). Did the mosasaurs spend most of their time in the deep oceans? Or did they live near the surface? How fast and how far did they swim? Since we cannot travel back in time to observe the animals in action, scientists have to use ingenious techniques to bring them back to life.

Consider the question of how fast mosasaurs could swim. Could they engage in sustained, fast swimming over long distances in the open ocean? Massare (1988) showed that mosasaurs probably could not swim very efficiently. She began by calculating the *fineness ratios* of different kinds of Mesozoic marine reptiles. The fineness ratio (F) is defined as the ratio of body length (L) to mean diameter (W).

$$F = L/W$$

Every marine animal has a fineness ratio, and one can easily estimate the fineness ratios of living creatures. For example, bottlenose dolphins have a fineness ratio of 4.4. Swordfish have a fineness ratio of 4.2. One can also estimate the fineness ratios of extinct creatures simply by measuring their skeletons.

The fineness ratio is related to the amount of drag that an animal must overcome during swimming. There are different kinds of drag, but the main one to think about is friction drag, which results from the flow of the water over the animal's body. If you could hold the body volume constant while increasing the fineness ratio, that would also increase the surface area – and the more surface area, the more drag. Scientists studying living organisms have shown that friction drag is minimized when the fineness ratio is about 4:5. The optimal range for fineness ratios is between 3 and 7. An animal with a fineness ratio lower than 3 or higher than 7 swims less efficiently, because drag increases considerably. If you have two animals, both swimming at the same speed and for the same length of time, the one that experiences greater friction drag will have to expend proportionally more energy to keep up. It's no coincidence that dolphins and swordfish have fineness ratios well within the optimum range, since both species engage in sustained fast swimming. Many other modern day swimmers, such as crocodiles and eels, have fineness ratios that are well outside the optimum range.

What about the mosasaurs? They also had fineness ratios well outside the optimum range – some in the neighborhood of 10 or 11. Massare's conclusion: mosasaurs would have been fairly lousy long-distance speed swimmers. This biomechanical finding might have some implications concerning what they ate and how they lived. Massare argues that they were probably ambush predators, and that their body type would have been better adapted for burst swimming and quick, lunging attacks.

Now contrast Massare's work on mosasaurs with an example of evolutionary paleontology. Many of the questions that evolutionary paleontologists work on have to do with extinction. Extinctions do not seem to be completely random; some species have a higher probability of becoming extinct than others. So what sorts of things might influence a given species' probability of becoming extinct? Incidentally, this is a question that interests conservation biologists as well as paleontologists, but the fossil record gives paleontologists a unique way of investigating it. One thing that might play a role here is the relative abundance of different species. Intuitively, it would seem like numbers can make a difference to extinction risk: a species that is rare would seem to be at greater risk of extinction, whereas a more abundant species seems more insulated against extinction. Indeed, at first glance, it seems like the relationship between abundance and species risk ought to be linear: the more a species increases in number, the more its extinction risk is reduced. After all, extinction is what happens when the population size falls to zero. So the further away from zero a given species is, the lower its risk of extinction.

In one recent study, Simpson and Harnik (2009) set out to test this idea that the relationship between species abundance and extinction risk is linear. Notice, though, that this is not the kind of claim that one can test by going out into the field and hunting for a particular fossil. Instead, Simpson and Harnik used an approach that was pioneered in the late 1970s and early 1980s by Jack Sepkoski, a paleontologist who was based at the University of Chicago (Ruse 1999). Sepkoski was the first scientist to use large computer databases as a tool for paleontological research. He found that if you have a database with detailed information about thousands of fossil specimens – all of which had been collected and painstakingly described by earlier scientists – then you can use the database to search for interesting patterns in the fossil record, a technique that Michael Ruse has aptly termed "crunching the fossils." Simpson and Harnik took advantage of the Paleobiology Database (PBDB), a publicly accessible and constantly growing reservoir of information about fossil collections from around the world. Simpson and Harnik focused on marine bivalves over the last 250 million years. Although many people immediately think of dinosaurs when they think of paleontology, evolutionary paleontologists tend to focus more on marine invertebrates, just because they leave behind lots and lots of fossils. To give an idea of the scope of Simpson and Harnik's study, they trolled through 1,631 different collections of fossils around the world, and together those collections housed 7,169,465 fossils.

What Simpson and Harnik found was highly counterintuitive. Instead of a linear relationship between species abundance and extinction risk, they found the U-shaped curve depicted in Figure 1.2. Extinction risk does increase with rarity. But at the same time, species that are *too abundant* also seem to have an increased risk of extinction. Much of Simpson and Harnik's work is devoted to analyzing possible sources of error and bias, just to make sure that their result is not a statistical illusion. The big question at the end, of course, is why superabundance comes with increased extinction risk. What is the cause of this "anomalous yet persistent" pattern that seems to be showing up in the fossil record of marine bivalves? They decline to speculate, preferring instead to leave the "why" question for future research.

In both of these case studies, the scientists employ rigorous quantitative techniques. Massare uses biomechanical modeling to help answer questions

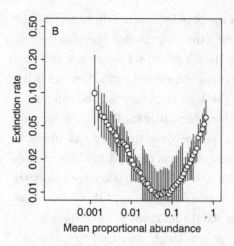

Figure 1.2 Simpson and Harnik's (2009) U-shaped curve. The proportional abundance of a genus is the ratio of its abundance to the total sample size of a fossil collection. The mean proportional abundance for a genus is arrived at by taking the average across many collections. For discussion of the details of calculating the extinction rate, see Simpson and Harnik (2009, p. 633). As the mean proportional abundance increases, the extinction rate drops, and then increases again. Reprinted with permission of the Paleontological Society.

about mosasaur swimming behavior, whereas Simpson and Harnik use sophisticated statistical tests to check the hypothesis that the relationship between abundance and extinction risk is linear. Yet Massare is investigating questions about mosasaurs, while Simpson and Harnik are investigating questions about evolution. These different emphases help explain why the scientists work with such different data sets. Massare needs only a few mosasaur skeletons to supply measurements of fineness ratios, but Simpson and Harnik need to look at millions of fossils in order to discern larger evolutionary patterns.

The paleobiological revolution

In the early 1970s, a number of paleontologists had grown dissatisfied with the position of their discipline on the sidelines of evolutionary theory. They set out to change things. The best-known member of that group was Stephen Jay Gould, but it included a number of other leading scientists as well: J. John Sepkoski, Jr., David Raup, Thomas J.W. Schopf, Steven Stanley, Elisabeth Vrba, Niles Eldredge, and others. These ambitious scientists helped to bring about what historian of science David Sepkoski has called "the paleobiological revolution" (Sepkoski 2009a). Not that there was nothing interesting going on in paleontology before the 1970s. Many of the scientists just mentioned took inspiration from the work of George Gaylord Simpson, a paleontologist of the previous generation whose classic text, *Tempo and Mode in Evolution* (1944), helped to clarify basic questions and concepts that would shape much of the research that paleontologists would go on to do. Simpson sought to show how paleontology fitted into what is known as the "modern synthesis" of Darwin's evolutionary theory with classical genetics – that is, how it fitted in with the latest and best thinking about evolution at the time.

Though influenced by the work of Simpson and other mid-century scientists, the "revolutionaries" of the 1970s went further. This new generation of scientists aimed to shake things up in at least seven different ways. Evolutionary paleontology as we know it today is largely a result of their efforts. I'll try to capture the spirit of the young Turks of the 1970s and 1980s with seven revolutionary slogans.

"Paleontology has more to contribute to biology than to geology." Paleontology has always occupied an awkward position between these two sciences. The study of fossils has long played an important role in geology, in part because understanding fossils is helpful for identifying and dating types of rocks. The 1970s revolutionaries sought to move their field closer and closer to biology, and one way in which they tried to do that was to show that they had something to say about evolutionary theory. They founded a new scientific journal in 1975, called *Paleobiology*, in order to provide an outlet for studies of the fossil record that addressed questions about evolution. Some started using the term "paleobiology" to describe their work. This was a way of signaling that the game had changed; paleontologists were now contributing to evolutionary theory.

"Study fossils in bulk – individual specimens don't tell you much about evolution." A single fossil find can sometimes have the effect of dramatically changing our views about prehistoric life. One such find was John Horner's discovery of dinosaur nesting sites that forced a rethinking of dinosaur social life (Horner and Makela 1979). For all the excitement that attends those discoveries, it remains true that the best way to investigate general questions about how evolution works is to study huge collections of fossils in order to identify patterns. Jack Sepkoski's use of large databases represents an important innovation in this direction. Rather than seeking out the exciting individual

fossil finds, these scientists turned to statistical techniques for describing and analyzing larger patterns in the fossil record.

"*Paleontology needs theories.*" In the physical sciences, there has long been a distinction, though not a sharp one, between two sorts of activities. On the one hand, there is the activity of making novel theoretical contributions, of coming up with new ideas and new ways of seeing things. Then there is the more workaday activity of designing and overseeing experiments, taking measurements, collecting data, building new experimental apparatus, and so on. Although both of these activities are obviously important, the more theoretical work has long been accorded higher prestige. (If you name a few highly regarded "scientific geniuses" off the top of your head – Isaac Newton, Charles Darwin, Albert Einstein – one thing they all had in common was that they all did theoretical work.) Up until the 1970s, paleontology had no theory of its own. There was, of course, evolutionary theory, which is based on the work of Darwin and Gregor Mendel, but paleontology had not (yet) contributed much to the shaping of that theory. The new paleobiologists set out to show that their science had a theoretical contribution to make. The 1970s and 1980s saw the development of new theoretical approaches, especially punctuated equilibria (hereafter PE) and species selection. Paleontologists began to argue that the fossil record requires the revision and/or expansion of traditional evolutionary theory.

"*If you can't experiment, then simulate.*" Paleontologists are hampered to some extent by their inability to perform direct experiments on the past (Turner 2007). However, in the early 1970s, a group of scientists met at the Marine Biological Laboratory in Woods Hole, Massachusetts, and developed the first computer simulation of large-scale evolutionary processes (Huss 2009). It became possible to test ideas about evolution by running virtual experiments. These new modeling techniques had a major impact on paleontology, especially with respect to scientists' thinking about the role of chance and randomness in evolution. The computer models made it possible to study chance in a way that no one had ever done before.

"*Don't just assume that the fossil record is incomplete; analyze the incompleteness.*" When Darwin first published the *Origin of Species* in 1859, he had to explain away the fact that no one had found any intermediate forms in the fossil record – no "missing links" between older species and newer ones. So he argued that the geological record is incomplete, and that geological processes erase a great deal of information about prehistoric life. This move, as much as anything else, helped to nudge paleontology to the sidelines of evolutionary theory. If the fossil record is so incomplete, what does it have to teach us about evolution? In the 1970s and 1980s, paleontologists began to challenge this assumption of incompleteness in various ways. The theory of PE represented one kind of challenge (see Chapters 2 and 3). But scientists also insisted on making the incompleteness of the fossil record itself into an object of study. Once we understand some of the sampling biases in the fossil record, we can correct for them. Philosophers use the term "epistemology" to refer to the theory of knowledge. These scientists wanted to make the epistemology of paleontology a part of paleontology itself. Studying the incompleteness of the fossil record is a way of studying the limits of our knowledge of the past.

"Resist reductionism." In the mid twentieth century, once the modern synthesis in evolutionary biology was well established, many scientists took a reductionist view of the relationship between *macroevolution* and *microevolution*. Microevolution consists of changes in gene frequencies in populations – the sorts of changes that modern evolutionary theory describes and explains so well. Macroevolution is any change that occurs above the species level, such as increasing biodiversity (that is, increasing numbers of species). The mid-century reductionists tended to think that macroevolution is "nothing but" microevolution. If you could understand the causes of all the microevolutionary changes taking place over vast sweeps of evolutionary time, then you would know all there is to know about macroevolution. The new generation of paleontologists in the 1970s and 1980s launched a sustained attack on this reductionist outlook. They argued that there are some macroevolutionary patterns and trends that could not be mere by-products of changes taking place at the microevolutionary level. Some pushed for a newly expanded hierarchical view of evolution, which would allow for irreducible mechanisms operating at the macro-level. The centerpiece of this new hierarchical view was the concept of species selection (Chapters 4 and 5).

"Don't shy away from raising big questions about evolution." How important is natural selection as a cause of evolution? To what degree is the course of evolutionary history a matter of chance? Is evolution progressive? Was the evolution of intelligent, language-using, tool-using, creatures like us inevitable in the end, or do we owe our existence to historical accidents? Some of the new paleobiologists thought that the fossil record holds the key to answering some of these questions.

Hopefully this sketch captures some of the spirit of the new paleontology that emerged in the 1970s and 1980s. Not all of the scientists involved in these new developments would have endorsed all of these slogans, and they have not always agreed about what the new science should look like. David Sepkoski (2005; 2009a) has written about this little-known scientific revolution in much greater detail, from an historian's perspective. In many respects, the paleobiological revolution represented a move in the direction of evolutionary paleontology. In saying this, I don't mean to suggest that the scientists necessarily wanted to move away from organismic paleontology. The study of particular kinds of prehistoric life remains an important and lively part of paleontology. Questions about evolution often turn out to have connections with questions that call for organismic reconstruction. It would be a mistake to think of the paleobiological revolution as a revolution against organismic reconstruction. Rather, the major players of the "revolutionary" period – the 1970s and 1980s – sought to establish evolutionary paleontology as an important part of the field, without necessarily edging out other things.

Perhaps more than any other scientific field, paleontology today has a complicated relationship with its public image. For one thing, paleontology has a special role to play as a gateway science. Many young people first get excited about science through reading dinosaur books, visiting natural history museums, or watching dinosaur specials on television. Journalists are often quick to report on the latest dinosaur discoveries, and a number of dinosaur scientists, including Robert Bakker and John Horner, have written popular books about their work. My impression is that many paleontologists are quite happy that their field has such a high profile in the wider culture. This high profile does have a downside, however. For one thing, it's natural that when most people think of paleontology, they think of organismic reconstruction in general, and of dinosaur science in particular. (I should add that not all dinosaur science involves organismic reconstruction, but much of it does.) Evolutionary paleontology, with its more theoretical bent, remains less well known. In many ways, dinosaur science remains paleontology's public face, even though much of the action over the past forty years – not *all* of the action, but a lot of it – has occurred in evolutionary paleontology.

Paleontology's high public profile has another downside as well. Within the natural sciences, and possibly within academia more broadly, there is a widespread prejudice against those who write for broader, non-academic audiences. The general thinking behind this prejudice seems to be something like the following: "If non-specialists can understand it without too much effort, then it must not be that serious." Or maybe the thinking is like this: "Why would you go to such effort to reach a wider audience, when you could be doing more serious, high-level research?" I certainly see this attitude sometimes in my own discipline, philosophy. Here I just want to point out one thing that it means for paleontology: paleontology's high public profile – and especially its appeal to kids – sometimes fuels the perception that it is not quite as weighty or as rigorous or as technical as, say, the rest of evolutionary biology. That is precisely the perception that the paleobiological revolution was a revolution against. The career of Stephen Jay Gould embodies the irony of this situation. On the one hand, Gould was a leading figure in the paleobiological revolution. He contributed to several of the main theoretical innovations in evolutionary paleontology, including PE and species selection. On the other hand, he was also one of the greatest essayists and popular science writers of all time. Although he wrote numerous technical papers, some of which I'll discuss in what follows, he also defended his controversial views about evolution in his essays and popular books. This has made him vulnerable to the charge that his ideas and arguments are less rigorous and technical than one would like. Ironically, one of Gould's lifelong aims as a scientist was to win a place for paleontology as a rigorous, technical discipline.

With a few exceptions, philosophers of science have not engaged much with the recent developments in evolutionary paleontology. This book is meant to serve as a (mostly) non-partisan guide, with a strong philosophical slant, to some of the big ideas and questions about evolution that came out of the paleobiological revolution.

Philosophical questions in evolutionary paleobiology

In what follows, we'll examine some of the main issues and arguments of contemporary evolutionary paleontology. The book is divided roughly into four parts, each of which engages with one major theme of evolutionary paleontology:

- PE (Chapters 2 and 3)
- Species selection and the hierarchical expansion of evolutionary theory (Chapters 4 and 5)
- The study of large-scale directional trends in evolutionary history (Chapters 6 and 7)
- Debates about the role of contingency and chance in evolution (Chapters 8 and 9).

In each of these four areas, the discussion is heavy on the science, with plenty of examples of the kinds of work that paleontologists do today. But I explore these issues from the perspective of contemporary philosophy of biology and philosophy of science, with philosophical questions and concerns in mind.

It's helpful to contrast two different approaches to the philosophy of science. The *philosophy-first approach* begins with large, overarching questions about science in general, such as:

What is science? How does it differ from various forms of pseudoscience?
What is causation?
Is science making progress?
What is the aim of science?
What does it mean for evidence to count for or against a scientific theory?
Should we believe what our theories say, even about things that no one can
  observe?
Are there laws of nature?
What does it mean for one theory to be reduced to another?
What counts as a good scientific explanation?

We might also call this *top-down* philosophy of science. When top-down philosophers of science discuss real science, they usually try to use examples of scientific research to illustrate or support a philosophical answer to one of the above questions. The inquiry usually begins with these high-altitude questions and then incorporates examples of real scientific research on an as-needed basis.

Contrast this with the *science-first approach*, or what you might call bottom-up philosophy of science. Bottom-up philosophy of science typically begins with a close examination of the scientific practices in this or that area of research. Philosophers of science who go this route usually try to learn as much as they can about developments on the ground in the field they want to study, and they try to get wrapped up in some of the same questions that occupy the scientists. However, bottom-up philosophy of science, when done well, is not merely descriptive. The challenge is to begin with the science and then gradually work one's way up into philosophical territory, usually by following up on conceptual or normative questions that arise during the course of scientific research. Those who take the bottom-up approach draw on the traditional methods and concepts of philosophy on an as-needed basis.

The bottom-up philosopher of science should also be open to exploring meta-level questions – that is, questions about the particular area of science under scrutiny – along the way.

This book may serve as an illustration of the bottom-up philosophy of science. I won't try to offer an elaborate defense of this approach here, in part because this kind of methodological decision can only be justified by its fruits. My general strategy is to begin with a careful look at recent work in evolutionary paleontology and then follow up on the more philosophical questions that inevitably occur. To give you a sense for how this procedure might work, consider the following examples of philosophical questions suggested by each of the four themes explored in the following.

(1) *PE and the theory-ladenness of observation.* When Eldredge and Gould (1972) first introduced their theory of PE, they quite deliberately referred to the work of philosophers of science, such as N.R. Hanson and Thomas Kuhn, who had argued that all observation in science is shaped by the theories that scientists already hold. Eldredge and Gould argued that paleontologists' interpretations of the fossil record were being shaped by background assumptions about evolution – namely, the assumption of gradualism. So you can't really understand the debate about PE without asking whether observation is, in general, theory-laden. To understand the science, you need to do some philosophy (see Chapter 2).

(2) *Species selection and reductionism.* The scientists who defended species selection thought that one biological species might have traits that make it fitter than others. In the above example, Simpson and Harnik were interested in seeing whether abundance might be such a trait. In one sense, though, a species is *nothing but* a collection of individual organisms. For that reason, it seems like any trait a species has must be reducible to the traits of its individual members. If the features of species are reducible in this way, how could species selection be a distinct evolutionary mechanism? Fascinatingly, much of the debate about species selection is really a debate about reductionism. Once again, to understand the science, it is necessary to do some philosophy (Chapter 5).

(3) *Trends and progress.* Evolutionary paleontologists try to identify large-scale directional trends in the fossil record. One thing that motivates this work, in some cases, is a concern about evolutionary progress. The concept of progress is an example of what philosophers call a "thick" concept. It is

partly descriptive, and partly evaluative. Paleontologists do the descriptive work, but it's impossible to put that descriptive work in the proper context without first establishing what one should say about the notion of evolutionary progress – a philosophical task. Doubts about the whole idea of progress have motivated some scientists to voice skepticism about the study of directional trends. You need to do some philosophy in order to obtain a clear view of what is at stake in the scientific study of trends (Chapter 6).

(4) *Historical contingency.* Gould argued that evolutionary history is highly contingent. If you could somehow rewind the tape of evolution and play it back, things would work out very differently. In opposition to Gould, another paleontologist, Simon Conway Morris, has argued that convergence, rather than contingency, is the hallmark of evolution, and that evolution often leads to the same outcomes from different starting points. This disagreement has largely to do with "what if" scenarios. For example, what if the dinosaurs had not become extinct 65 million years ago. Would they eventually have evolved into intelligent creatures like us? It's not easy to see how to test claims about "what if" scenarios. Do such claims have any legitimate role to play in science? That's a philosophical question (Chapter 8).

In bottom-up philosophy of science, where does the discussion of science end and the discussion of philosophy begin? Sometimes it may be easy to tell when one has crossed the frontier from natural science into philosophy, from the empirical into the realm of the conceptual and the normative. Sometimes, though, the frontier is vague and difficult to make out; questions of theoretical definition (e.g., What is species selection?) interweave with questions about the world (e.g., Does species selection ever occur in nature?) in a way that often makes it difficult to tell whether one is talking about ideas or about things in the world. Natural science and philosophy are like two countries on a map with a common border that is well-marked in some places, disputed in others, and in still other places completely undefined. It's possible to start in the middle of the one country and head in the general direction of the other one without being able to say, or even much caring, when one has crossed the border.

What about the future of paleontology? Since the early 1990s, new developments in molecular biology have begun to have an impact on the study of prehistoric life. For example, scientists can now sequence the genomes of recently extinct creatures, such as the cave bear and the woolly mammoth. The proteins and DNA of living creatures can also provide evidence concerning the timing of evolutionary events in the very distant past. Research on so-called "molecular clocks" has both challenged and supplemented work on the fossil record. In the last chapter of the book, I examine these new developments with an eye toward philosophical questions about the meanings of scientific terms, such as "fossil." I suggest that we can best understand this new work by thinking of it as changing the very meaning of "the fossil record." This, in turn, links up with the thought experiment with which I began. In order to imagine a world without fossils, we first have to be clear what we mean by "fossil," and that meaning is liable to change.

2　A new way of seeing the fossil record

In 1972, Stephen Jay Gould and Niles Eldredge published a paper entitled "Speciation and Punctuated Equilibria: an alternative to phyletic gradualism." This paper, which built upon earlier work by Eldredge (1971), sparked a debate about what, if anything, paleontology has to add to evolutionary theory. Even though much of the controversy has died down, a good understanding of that debate is crucial for anyone who wants to think about paleontology's contribution to evolutionary theory. In this chapter, I will trace some of the main arguments in the early discussion of Eldredge and Gould's idea, but I want to do so with three guiding questions in mind:

1. Over the years, Gould repeatedly presented PE as posing a challenge to traditional evolutionary thinking. How exactly does PE depart from a reductionist – or as Gould often called it, an "extrapolationist" – view of evolution?
   And what exactly are the phenomena, trends, or patterns that call for such
   a departure?
2. To what extent did philosophical ideas that were "in the air" in the early
   1970s – especially the ideas of Thomas Kuhn – influence the discussion of
   PE?
3. Since scientists cannot directly observe past evolutionary processes, and
   since they cannot experiment on large-scale evolutionary processes, how
   have they gone about trying to test PE?

There are plenty of other good discussions of PE out there (see, for example, Sterelny 1992; Sterelny 2001, Chapter 8; Prothero 1992; Princehouse 2009; Sepkoski 2009b). The second of the three questions above gives my approach in this chapter a somewhat different emphasis from most of what has gone before. I suggest that paleontology has been deeply affected by philosophical ideas, for better or for worse.

Speciation

Eldredge and Gould (1972) presented their PE model as if it were merely a consequence of thinking about speciation in a certain way. So in order to understand PE, we need to think about speciation.

First, some definitions: biologists typically distinguish between two kinds of speciation, cladogenetic *vs.* anagenetic. Cladogenetic speciation occurs when a species or lineage branches and gives rise to two new species. ("Cladogenesis" just means branching.) Anagenetic speciation occurs without any branching. This could happen if a lineage goes through so much evolutionary change during a given time interval that we decide to say that it has actually changed from one species into another. The idea of anagenetic speciation is controversial. Some scientists prefer to define "species" in a way that rules out the possibility of anagenetic speciation. For example, the phylogenetic species concept (which is just one of several ways of thinking about species) takes a species to be a lineage that originates at one of the most recent branching points in the tree of life. That rules out anagenetic speciation by definition. For present purposes, we can just set that issue to one side and focus on cladogenetic speciation.

One of the central claims of Darwin's evolutionary theory is simply that cladogenetic speciation occurs. This claim is often stated in the form of a thesis about common ancestry: pick any two existing species, and they will have descended, with modifications along the way, from an earlier common ancestor. When Darwin first defended this idea publicly in 1859, the main rival was the theory that each species has a distinct origin in the past. Those who believed that each species was created by God in a distinct historical event were committed to denying the reality of cladogenesis. So we might ask a couple of tough questions of Darwin here: first, what is the empirical evidence that cladogenesis actually occurs in nature? And second, how does cladogenesis work?

Let's start with the second question. Darwin conceived of cladogenesis as a gradual process that is mostly driven along by the mechanism of natural selection. He famously devoted big chunks of the early chapters of the *Origin of Species* to discussion of animal breeding. Breeders had in many cases started out with an ancestral population and, over the course of many generations, succeeded in producing distinct varieties. So, for example, beagles and greyhounds are not distinct species, but they certainly are distinct canine varieties that have descended, with modification, from a common ancestral type. Darwin thought that speciation must occur in much the same way "out there" in nature. We often observe that biological populations contain different variants. Over long stretches of time, these sub-populations could grow more and more distinct with each generation. Today, biologists refer to this type of gradual process as *sympatric speciation*. "Sympatric" here means occupying the same geographical area; the idea is that speciation can occur within the ancestral range of a species. To give an example: consider a species of birds that feeds on two types of seeds. That species contains two variants with slightly different beak shapes, where each shape type is slightly better adapted for eating one of the two types of seeds. Over time, these two variants specialize more and more, and their beak shapes become more and more different. Eventually we get two distinct species, each of which is ecologically specialized to exploit one or the other seed type. This speciation process takes place very slowly, and it takes place within the range of the ancestral species.

Notice that the mechanisms involved in this sort of speciation process are essentially the same as those that cause trends in trait (or gene) frequencies in populations. Natural selection, perhaps, is the main one, but we might also include the other "forces" that can give rise to microevolutionary change: drift, mutation, and so on.

Now to return to the first of the two questions posed earlier: how do we know that cladogenetic speciation really occurs in nature? How can we test this idea? Darwin knew that simple observation can take us a long way, for we can observe the microevolutionary processes at work in natural populations. All we have to do is to imagine those processes taking place over long stretches of time. In other words, all we have to do is extrapolate backwards. In addition, Darwin saw that the hypothesis that speciation really occurs can do a huge amount of explanatory work in biology. To give just one quick and well-worn example, it explains the occurrence of homologies, or otherwise puzzling structural similarities between two distinct species. Why do African and Asian elephants both have tusks and trunks? These similarities are easily explained on the supposition that the two species descended from a recent common ancestor. That means, in turn, that the common ancestor must have gone through a process of cladogenetic speciation. But Darwin also knew that there was an actual prediction that could test his claims about speciation and common descent.

Transitional forms

If cladogenetic speciation really occurs in the gradual way that Darwin suggested, then we should expect to see many transitional forms in the fossil record. We ought to see at least some series in the fossil record which document evolution's passage from point A to point B. This expectation of transitional forms in the fossil record seems to follow from Darwin's assumption that speciation is a gradual process. In 1859, however, Darwin fretted because he knew that up to that time, no one had found any clear examples of such transitional forms:

> The main cause, however, of innumerable intermediate links not now
> occurring everywhere throughout nature depends on the very process of
> natural selection, through which new varieties continually take the places of
> and exterminate their parent-forms. But just in proportion as this process of
> extermination has acted on an enormous scale, so must the number of
> intermediate varieties, which have formerly existed on the earth, be truly
> enormous. Why then is not every geological formation and every stratum full
> of such intermediate links? Geology assuredly does not reveal any such finely
> graduated organic chain; and this, perhaps, is the most obvious and gravest
> objection which can be urged against my theory. The explanation lies, as I
> believe, in the extreme imperfection of the geological record. (1859/1964,
> pp. 279–280)

The reason we do not see many transitional or intermediate forms living today, alongside their descendants, has to do with natural selection. When gradual cladogenetic speciation occurs, the ancestral population splits into two offspring populations, each of which evolves on its own new trajectory. At a certain point – and it will always be somewhat arbitrary where exactly we draw the line – the ancestral species ceases to exist, and we have two new offspring species in its place. So it should come as no surprise that there are no living transitional forms. What about transitional forms in the fossil record?

Many philosophers and scientists agree that the very best kind of empirical evidence that one can acquire in science is *novel predictive success*. This occurs when a theory makes a novel prediction that comes out true. But what counts as a "novel" prediction? Philosophers disagree among themselves about how best to characterize predictive novelty, but most agree that the theory from which the prediction is derived must not be tailored to generate that very prediction (Leplin 1997). In other words, the predicted result must not be among

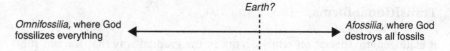

Figure 2.1 The continuum of completeness.

the observational data that the theory was originally designed to explain. Another important condition is that no other theory may make the same prediction: if two rival theories predict exactly the same result, then observing that result won't help discriminate between them.

Darwin's prediction of the existence of intermediate forms in the fossil record meets both of these requirements. To begin with, the rival theory that God created each living species in distinct historical interventions does not lead one to expect that there should be intermediate forms in the fossil record. The discovery of intermediate forms might not disprove the creationist theory, because there might be clever ways for creationists to accommodate the new discovery. The point is just that there is nothing about creationism which leads us to expect, going in, that there should be intermediate forms. What's more, Darwin did not himself know whether any transitional fossils would actually be discovered. Scientists in the meantime have discovered lots of them, beginning with *Archaeopteryx* in the 1860s. *Archaeopteryx* is the dazzling fossil first found in the Solnhofen quarry in Bavaria that has both birdlike and reptile-like features – at the time, a plausible transitional form between dinosaurs and birds. (The story about how birds evolved from dinosaurs has become much more complicated.) The point is that Darwin worked out his theory without access to these observational data. His evolutionary theory could not have been tailored to accommodate findings that he knew nothing about.

Darwin's gradualist picture of speciation suggests that there should be an awful lot of transitional fossils out there waiting to be found. The failure up to that point to find them would seem like a problem. To address it, Darwin makes a notorious defensive move: since the fossil record is so incomplete, we should not be so surprised at the dearth of intermediate forms, even if evolutionary theory is largely true.

Darwin correctly claims that the fossil record at best gives us a patchy, incomplete picture of the history of life on Earth. It may be helpful to imagine a continuum of historical completeness, with Afossilia at one end (see Figure 2.1). At the other end of the continuum is another planet – call it

Omnifossilia – whose rocks are crammed full of fossils. Omnifossilia contains a (nearly) perfectly complete evolutionary record. We need not suppose that the remains of every single organism that ever lived on Omnifossilia was fossilized. Instead, suppose that God steps in once per century and (somehow) fossilizes one pair of organisms from every single population on Omnifossilia. Omnifossilia is the geological version of Noah's ark, where each type of creature is preserved for all posterity. We can also imagine that on Omnifossilia, God (somehow) includes fossils that contain information about the soft parts of organisms as well as the teeth and bones. (Incidentally, this thought experiment about Omnifossilia suggests an interesting reply to those who may wonder whether God placed the fossils in the rocks to test our faith in scripture. Think of all the *other* things God could have placed in the rocks to tempt and confound our meager intellects!)

Where does our own planet fall along the continuum of completeness? It probably belongs closer to the Afossilian end of the spectrum. Just how close to Afossilia we are is an empirical question that we shouldn't try to settle here. But there is no doubt that our own fossil record is both massively incomplete and systematically biased. Some of the biases can be brought out by reflecting that Earth's position along the continuum of completeness may depend on which sorts of organisms one is interested in. If you are focusing on marine invertebrates living in shallow water, Earth's fossil record is pretty good. Many marine invertebrates have hard parts (especially shells) that fossilize readily when buried by sediment on the seafloor. This is one reason why paleontologists have studied marine invertebrates so intensively. In many ways, they represent our best shot at getting an accurate picture of large-scale evolutionary patterns. Our fossil record for many other groups is less impressive.

As it happens, scientists have found numerous examples of transitional forms in the fossil record. In 2006, a team of scientists published a description of a fossil specimen they had found on Ellesmere Island in the Canadian arctic (Daeschler, Shubin, and Jenkins 2006). This creature, which they named *Tiktaalik*, lived in the earliest part of the late Devonian period, about 380 million years ago, and it is a clear example of a transitional form between lobe-finned fishes and the earliest land-dwelling tetrapods. It has many fishlike features. For example, its limbs end in ray fins, rather than feet with digits. But in several respects, it looks like a fish-on-the-way-to-becoming-a-tetrapod.

One transitional feature of *Tiktaalik* has to do with the bones around the gills. It has a wider spiracle, or gill slit, than other lobe-finned fishes, and indeed its spiracle bears a closer resemblance to the bones making up the inner ears of tetrapods that lived just a few million years later. In addition, close study of the animal's pectoral fins suggests that they could have been used to support the animal's weight if it ever wanted to walk in shallow water. While a single fossil specimen cannot tell the whole story about the evolution of the earliest land-dwelling tetrapods, it does represent a novel predictive success for evolutionary theory. Although creationists can perhaps give *ad hoc* explanations of these results, there is nothing in the creationist account of the origin of species that gives us any reason to expect to find fossils such as *Tiktaalik*.

Still, it's worth pointing out that transitional fossils remain comparatively rare, which is why evolutionists celebrate finding one like *Tiktaalik*. The unearthing of a fossil such as *Tiktaalik* only has the effect of reducing the size of the gaps in our knowledge of evolutionary sequences. There is still quite a lot of evolutionary distance (and several million years of geological time) between *Tiktaalik* and the earliest *bona fide* tetrapods, and we don't yet have any direct fossil evidence of the animals which might have occupied that gap. Since we do not live on Omnifossilia, scientists will never be able to eliminate the gaps in the fossil record. The discovery of transitional forms just makes those gaps smaller.

When Eldredge and Gould first introduced their theory of PE, they meant to challenge the traditional way of thinking about transitional fossils and gaps in the fossil record.

> Paleontology's view of speciation has been dominated by the picture of "phyletic gradualism." It holds that new species arise from the slow and steady transformation of entire populations. Under its influence, we seek unbroken fossil series linking two forms by insensible gradation as the only complete mirror of Darwinian processes; we ascribe all breaks to imperfections in the record. (1972, p. 84)

In other words, paleontologists approach the fossil record with certain assumptions and expectations. They all "know" going in that what they should see there – indeed, what they would see there, if they were on Omnifossilia – is an "unbroken fossil series linking two forms by insensible gradation." When they don't see that, the record itself is to blame. But what if the apparent gaps

Figure 2.2 The duck-rabbit.

in the record were not really gaps at all? What if Earth is more like Omni-fossilia than we have realized? Perhaps paleontologists have been looking at the fossil record through the wrong lenses, so to speak. Scientists thought all along that transitional forms are the signal while everything else is mere noise, but what if they had that backwards? What if the gaps in the fossil record actually contain information about how evolution works?

Gould's *Gestalt* shift

In the 1950s and 1960s, a number of important philosophers were deeply impressed with the findings of what was then known as *Gestalt* psychology. (The German term *Gestalt* means something close to "form," or "structure.") The *Gestalt* psychologists were interested in the ways in which people perceive things such as the duck-rabbit drawing (see Figure 2.2). Suppose that two people look at this drawing. One seems to see, and perhaps really does see, a duck; the other seems to see a rabbit. And yet, in a sense, what they are seeing is exactly the same thing. How should we describe what is going on here? And why is it not possible to see both the duck and the rabbit simultaneously? These questions were very much in the air in the 1950s and 1960s, and a number of important philosophers, from Wittgenstein (1953/1973) to N.R. Hanson (1958) to Thomas Kuhn (1962/1996), engaged with them in a sustained way. Imagine the experience of a person who sees this drawing as a duck and then suddenly, for the first time, has the experience of seeing it as a rabbit. That person, we might say, has undergone a *Gestalt* shift.

Eldredge and Gould (1972) argued that entrenched modes of thinking about evolution had caused scientists to see the fossil record a certain way, much as someone who expects to see a duck will see just that when looking at

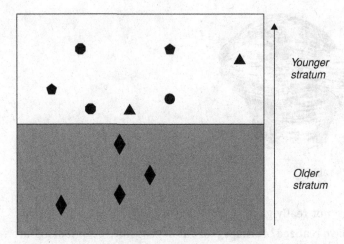

Figure 2.3 The fossil record. Only one species (the diamond) occurs in the lower stratum. It disappears and is replaced by new species (the pentagons, triangles and circles). There are no transitional forms.

the duck-rabbit. They suggested that the fossil record resembles the duck-rabbit. In Figure 2.3, in the lower (older) stratum, a single species is well represented. In the upper (younger) stratum, that older species has given way rather abruptly to three newer species, which seem to be descended from it. If you approach this diagram with Darwinian gradualist predilections, then you will assume that the evolutionary process by which the ancestral species (the diamond) gave rise to its descendants (the triangle, circle, and pentagon) must have been rather slow and steady. There must also be numerous intermediate forms between the older and the younger species. The break in the fossil record is merely an illusion due to the incompleteness of that record. This might be an instance of what paleontologists and geologists call *stratigraphic incompleteness* (Kemp 1999, pp. 85ff). We should not assume that the rocks we are looking at in this case were formed at a steady rate. Perhaps the break between the two strata corresponds to quite a long time interval during which, for whatever reason, no new sediments were deposited at this spot. That could happen, for example, if a major river shifted its course, or if a shallow lake dried up for a long period. And perhaps there was a lot of gradual evolution going on during that "missing" time interval.

What if the fossil record as depicted in Figure 2.3 is more or less complete? The lower stratum tells the story of a time interval during which very little evolutionary change took place – a period of relative stasis, or if you will, a period of equilibrium. Then there is a much shorter interval during which

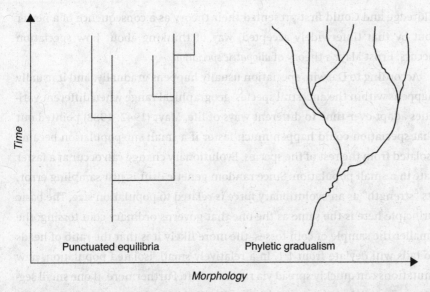

Figure 2.4 PE *vs.* phyletic gradualism. Notice that on the left-hand side, the evolving lineages make lateral moves through the morphospace that are nearly instantaneous. Once new lineages are formed, they persist for long periods with no morphological change.

a great deal of cladogenetic speciation occurred. This speciation must have taken place very rapidly, at least in geological terms. This period of rapid change is then followed by another stretch of evolutionary peace and quiet, as is reflected in the upper stratum. To look at the fossil record in this way is to undergo a significant *Gestalt* shift: the abrupt break that once seemed due to stratigraphic incompleteness now tells the story of rapid speciation and evolutionary change. The contrast between PE and gradualism is shown in Figure 2.4.

This *Gestalt* shift is the core of the theory of PE. However, understanding this basic move is just the beginning of the story. There are additional complexities, consequences, and ramifications to explore. Even if Eldredge and Gould were right that the fossil record is like a duck-rabbit drawing, why should we see things their way? And what difference does it make if we do?

How radical is PE?

The seminal Eldredge and Gould (1972) paper appears in a volume edited by another scientist whom we will encounter again later on, Thomas J.M. Schopf. Schopf originally asked Gould to contribute a paper on speciation. In fact,

Eldredge and Gould first presented their theory as a consequence of a newer (but by that time widely accepted) way of thinking about how speciation occurs: Ernst Mayr's theory of *allopatric speciation*.

According to Darwin, speciation usually happens gradually, and it usually happens within the ancestral species' geographical range when different varieties adapt over time to different ways of life. Mayr (1942; 1988) pointed out that speciation could happen much faster if a small sub-population became isolated from the rest of the species. Evolutionary change can occur at a faster rate in a small population. Since random genetic drift is just sampling error, its "strength" as an evolutionary force is related to population size. The basic principle here is the same as the one that governs ordinary coin tossing: the smaller the sample of coin tosses, the more likely it is that the ratio of heads to tails will deviate from 1:1. In a relatively small, isolated population, new mutations can quickly spread via random drift. Furthermore, if one small segment of a population becomes geographically isolated from the rest, there will probably also be important environmental differences between the habitat of that isolated fringe group and the habitat occupied by the rest of the species. It might not take very long (at least not long in geological terms) for natural selection to adapt that fringe population to its own distinct environment.

Mayr's model of allopatric speciation differs in important ways from Darwin's original (sympatric, gradualist) model. The allopatric speciation model has two consequences that Eldredge and Gould emphasized. First, it shows how cladogenetic speciation could happen much more rapidly than Darwin realized, even though it would still take many generations. And second, it has the consequence that the ancestral species and the newer offspring species do not live in the same place. Eldredge and Gould then asked: What would it mean for paleontology if speciation is usually allopatric? That would mean that when we approach the fossil record, we should expect to see new species show up in the record quite abruptly. And we should not expect to see clear series of fossil remains documenting the evolution of one species into another, in part because the newer species are usually evolving in some location other than where their ancestors lived. The allopatric model of speciation suggests that much of the phenotypic change that takes place during evolution happens in spurts during speciation events, when splinter groups from an ancestral population rapidly adapt to a new environment. Once a species gets established, it might not evolve much after that. So if allopatric speciation is the norm, we shouldn't expect to see too many transitional fossils. Instead, we should expect to see periods of stasis, in which species do not go through much evolutionary change at all, punctuated by episodes of speciation and rapid evolutionary change. Thus, in their original presentation of the theory, Eldredge and Gould were essentially scaling up the model of allopatric speciation.

To put all this another way: whether you are predisposed to see the fossil record as informative *vs.* incomplete depends on which type of speciation – sympatric *vs.* allopatric – you think is the norm. Eldredge and Gould use the model of allopatric speciation to try to induce the *Gestalt* shift in other paleontologists, and to get them to see the apparent gappiness of the fossil record as containing information about how evolution works.

How revolutionary is this idea? There are at least two senses in which the PE model goes against Darwin's own thinking about these matters. First, it suggests that allopatric speciation is the rule, and sympatric speciation the exception. And second, it suggests that Darwin was not entirely correct about what G.G. Simpson (1944) called the *tempo* of evolution. Darwin equated evolution with gradual, steady, cumulative change. Eldredge and Gould argued that evolution happens in fits and starts. Most evolutionary change occurs rather quickly during speciation episodes. Once species get established, they don't change very much. Darwin looms so large in evolutionary theory that any attempt to correct or even supplement his ideas can seem like a very big deal.

In other ways, though, the theory of PE is not too radical. In fact, whether we should even call it a scientific *theory*, as opposed to a *way of seeing* the fossil record, is open to question. To begin with, the model of allopatric speciation is perfectly compatible with Darwin's account of microevolutionary processes. Indeed, the allopatric model invokes natural selection as one potential cause of rapid evolutionary change during speciation episodes. If we ask which mode of speciation has been more prevalent during evolutionary history, that is really a question about the historical details. The central core of modern microevolutionary theory is compatible with any answer. Indeed, this is a good example of what philosopher of biology John Beatty (1995; 1997) calls a "relative significance debate" – a type of disagreement he thinks is extremely common in biology. This kind of debate occurs when biologists have different models which represent natural processes and things in different ways. The question, then, is which model is instantiated by the highest proportion of actual cases? No one denies that allopatric speciation sometimes occurs, and that sympatric speciation also occurs in nature. But which is more significant?

There is another sense in which PE might not turn out to be so radical. Recall from Chapter 1 that one of the big questions of evolutionary paleontology concerns the relationship between macroevolution and microevolution. Many scientists have favored reductionist views, according to which macroevolutionary patterns are mere by-products of changes taking place at the micro-level over very long stretches of time. Todd Grantham offers the following characterization of this reductionist outlook:

> [T]o say that macroevolution is *explanatorily* reducible means that microevolutionary theory can do all the explanatory work of macroevolutionary theory. To put this somewhat differently, combining an idealized lower-level theory with the set of background conditions would (in principle) allow us to explain any given macro event. (2007, p. 76)

The explanatory reductionist, in Grantham's sense, holds that all the macro-level patterns and events are fully explicable in terms of microevolutionary events and processes.

Kim Sterelny (2007) offers a slightly different way of thinking about this reductionistic outlook. He introduces the notion of a *minimalist model* of the relationship between macro- and microevolution. According to such a model:

> macroevolutionary patterns are direct reflections of microevolutionary change in local populations; they are reflections of changes of the kind we can observe, measure, and manipulate ... macroevolutionary patterns are nothing but local changes summed over vast sweeps of evolutionary time. (2007, p. 182)

Sterelny is careful to insist that we think of minimalism not so much as a doctrine about evolution, but rather as a family of evolutionary models. Minimal models of evolution are relatively parsimonious, so as a matter of methodology, we might want to start out working with minimal models and expand them only on an as-needed basis. As he puts it, "instead of thinking of minimalism as a doctrine to be defended or undermined, we should instead focus on identifying the range of cases for which minimal models are appropriate, and those cases in which these models need to be supplemented" (2007). Sterelny considers a variety of ways of supplementing minimal models, including some, such as species selection and macromutation, that we will consider later on.

The subtle differences between Grantham's and Sterelny's ways of framing the issue will not matter much for our present purposes. Both philosophers offer helpful ways of thinking about the significance of PE. Does PE suggest, contra explanatory reductionism, that any macroevolutionary phenomena resist explanation in microevolutionary terms? Alternatively, does PE give us a reason for expanding or enriching minimalist models of evolution? Perhaps surprisingly, I think the answer is no – at least not at first. Not only is PE compatible with a minimalist picture of evolution; it may even presuppose a minimalist picture.

Here's why. Minimalist models of evolution incorporate the following schema, and they idealize away from other factors that might complicate this simple picture.

Microevolutionary change + speciation → macroevolutionary patterns and trends

Eldredge and Gould initially posed no serious challenge to this formula. Instead, they use this very formula to help develop their case. If we think of a minimalist model of evolution as an account of microevolutionary change plus an account of speciation, then Eldredge and Gould deserve credit for pointing out that which macro-level patterns and trends you expect to see will depend on which minimal model of evolution you are working with. If you work with a model of sympatric speciation, as Darwin did, then you'll expect to see plenty of transitional forms in the fossil record and lots of series of fossil specimens documenting gradual evolutionary change. On the other hand, if you work with an allopatric model of speciation, you should expect periods of stasis punctuated by speciation and rapid evolutionary change. Even if PE, understood as a set of claims about evolutionary tempos, does not require a minimalist model of evolution, the argumentative strategy that Eldredge and Gould adopt in their early work certainly does seem to presuppose that macroevolution is "nothing but" microevolution.

The influence of Kuhn

Eldredge and Gould's original presentation of the theory of PE was marinated in the philosophical ideas of Thomas Kuhn's classic, *The Structure of Scientific Revolutions* (1962/1996). In fact, the influence of Kuhn's work is so profound, as I will document shortly, that PE deserves to be treated as an important case study of the ways in which ideas in the philosophy of science can influence the practice of science. Whether this influence has been a good or a bad thing in this case is a complicated question; much depends on what you make of Kuhn's work.

Over the years, many have speculated about the possible relationship between the theory of PE and Gould's supposed Marxist political leanings (Dusek 2003). I don't want to make too much of this myself. Gould himself did at one point acknowledge that Marxist thought was very much part of the intellectual milieu in which he was raised, and that this cultural context probably influenced his thinking about evolution in subtle ways – just as Victorian beliefs in progress and the desirability of economic competition surely influenced Darwin's thinking about evolution. But Gould's hackles often went up at the suggestion that his scientific ideas were ideologically motivated, and in fairness to him and Eldredge, we need to evaluate those ideas on their own merits. Still, it is impossible not to notice at least one glaring similarity between Gould's view of evolution and Marx's picture of historical change. Marx's picture is also one in which periods of relative calm and stability are "punctuated" by episodes of revolutionary change. Although he never used such terms, Marx clearly thought that human historical change exhibits a pattern of PE.

There is one glaring dissimilarity between Gould's view of prehistory and Marx's view of history. Marx, like many other nineteenth-century thinkers, confidently believed in human progress. He inherited from Hegel the idea that human history has a *telos* (or end) to which everything else is leading up, and for the sake of which everything else occurs. Gould, however, denied that evolutionary history has any such overarching *telos* or goal. Gould insisted over and over again, and in a variety of different ways, that evolution is heading nowhere in particular. On this question of the direction of history, it would not have been possible for Gould to put any more intellectual distance between himself and Marx.

On the other hand, there is scarcely any intellectual daylight between Gould and Thomas Kuhn. Like Gould (and Marx), Kuhn sees history as a series of punctuated equilibria. But unlike either Gould or Marx, the history that Kuhn focuses on is the history of science. The history of science, he argues, is characterized by many periods of relative calm and stability, which he calls *normal science*. During these periods of normal scientific research, scientists look to established *paradigms*, or past scientific achievements, for guidance. A paradigm provides an entire scientific community with an exemplar or a model for how to do good scientific work. The Newtonian paradigm in physics affords a good example of this. One of Newton's many achievements was to show how it is possible to give a unified account of terrestrial and celestial motion in terms of a single set of mechanical laws. This achievement gave later scientists a standard to live up to: namely, it showed them that they need to unify disparate phenomena by subsuming them under laws. At the same time, Kuhn stressed that paradigms need to be "open-ended," in the sense that they need to leave the community of researchers with unanswered questions to investigate, and unsolved puzzles to work on. During periods of normal science, researchers seldom question basic assumptions or construct new theories from scratch. Instead, they investigate localized questions that they are confident they can answer using the methods prescribed by the established paradigm.

According to Kuhn, these periods of normal science are occasionally punctuated by revolutionary episodes of scientific change. During the course of normal scientific research, anomalies gradually accumulate. An anomaly is a phenomenon or observation that doesn't fit too well with the picture of the world that is associated with the going paradigm. No single anomaly is ever sufficient to induce scientists to abandon their paradigm, but as anomalies grow in number, eventually scientists can enter a crisis mode where they begin to question the entrenched way of doing things. Sometimes, during these periods of crisis, an especially original thinker – usually a younger scientist or someone outside the establishment – develops an impressive new way of looking at things, a new achievement that can rival the existing paradigm. Scientists must decide where their loyalties lie, but if the new way of doing things wins enough converts, this period of revolutionary science can culminate with a shift to a new paradigm. It is an interesting question whether the paleobiological revolution involved a genuine Kuhnian paradigm shift (for discussion, see Ruse 2009). One thing is sure: Gould certainly took himself and the other scientists of his generation to be ushering in a new paleontological paradigm.

PE suggests that the pattern of change in evolutionary history strongly resembles what Kuhn takes to be the pattern of change in the history of science. Moreover, Kuhn argues that the history of science also has no overarching *telos* or goal. In the last chapter of *The Structure of Scientific Revolutions*, Kuhn draws a parallel between the history of science and evolutionary history – an analogy which must have influenced Gould and Eldredge. He acknowledges that the history of science "has been a process of evolution from primitive beginnings –

a process whose successive stages are characterized by an increasingly detailed and refined understanding of nature" (1962/1996, p. 170). But he also denies that science is developing toward anything, or in any particular direction. He denies, in other words, that science has any overarching goal or *telos*. One common view of science, held by many realist philosophers, is that science is gradually converging on the truth. Kuhn flatly rejects this view, partly on the grounds that successive paradigms are incommensurable. He does not think that there is any metric by which to compare earlier with later paradigms in order to determine which is closer to the truth. For anyone who finds this view tough to swallow, Kuhn points out that some have found Darwin's theory tough to swallow for the same reason:

> For many men the abolition of that teleological kind of evolution was the most significant and least palatable of Darwin's suggestions. The *Origin of Species* recognized no goal set either by God or nature. (1962/1996, pp. 171–172)

Gould also liked to emphasize the lack of any overarching goal or *telos* in biological evolution.

At the very least, it's clear that Gould's and Kuhn's historical visions have a lot in common. When Eldredge and Gould write that "science progresses more by the introduction of new world-views or 'pictures' than by the steady accumulation of information," they were just paraphrasing Kuhn (Eldredge and Gould 1972, p. 86). I want to go a bit further than this, however, and suggest that the influence of Kuhn also helps explain some otherwise puzzling aspects of the way in which Gould and Eldredge presented their theory. It may also help to explain how the controversy over PE escalated.

> The idea of PE is just as much a preconceived picture as that of phyletic gradualism. We readily admit our bias towards it and urge readers, in the ensuing discussion, to remember that our interpretations are as colored by our preconceptions as are the claims of the champions of phyletic gradualism by theirs. We merely reiterate: (1) that one must have some picture of speciation in mind, (2) that *the data of paleontology cannot decide which picture is more adequate*, and (3) that the picture of PE is more in accord with the process of speciation as understood by modern evolutionists. (Eldredge and Gould 1972, pp. 98–99, emphasis added)

Note some Kuhnian themes here: first, Gould and Eldredge refer to their own theory as a "preconceived picture" and say that their interpretations of the fossil record are "colored by our preconceptions." The thought seems to be that scientists will, in effect, see what they expect to see when they study the fossil record. Phyletic gradualism and PE just shape scientists' expectations in different ways. Their claim (1) suggests that it is impossible to look at the fossil record *without* some such preconceptions. You have to have some sort of view about speciation, and whatever that is, it will shape your expectations. The claim that stuck in the craw of many scientists was actually claim (2): there Eldredge and Gould seem to say that it is impossible to test the theory of PE using fossil evidence! Why would they say such a thing? Why, upon introducing a new theory, would any scientist ever want to claim that the theory is impossible to test?

This remark about the untestability of the model has puzzled many scientists. In fact, there have been many attempts to test the theory of PE, and in some of his later reminiscences on the development of the theory, Gould makes it abundantly clear that he thinks that PE has survived empirical testing. So why did he change his mind? And why would he and Eldredge ever have suggested that PE is untestable in the first place?

The answer is that this was a Kuhnian moment. In his seminal work, Kuhn had emphasized the *theory-ladenness of evidence*. The idea is that what you observe depends, in interesting ways, on your background theoretical commitments, or as Kuhn himself might say, on the paradigm that you are working under. This reverses the traditional view that theory should depend on evidence. The theory-ladenness of evidence can come in different strengths:

1. Your background theories might tell you where to look for evidence. For example, geological theories might tell you where to do your fieldwork if you want to study, say, the evolution of early land plants.
2. Your background theories might tell you why your evidence counts as evidence. For example, suppose you are looking at carbon isotope ratios in rocks in order to draw conclusions about the amount of photosynthetic activity in ancient oceans. You have to rely on background theories that explain how the carbon isotope ratios are related to photosynthetic activity.
3. Your background theories might tell you how to interpret what you see. Different scientists, with different theoretical commitments, might see the same thing but interpret it differently.
4. Your background theories might influence or even determine what you see. Different scientists, with different theoretical commitments, might look at the same object but (somehow) see different things.

Most philosophers of science today would agree that evidence is theory-laden in the first three of these respects. The fourth, much stronger claim, is more controversial. The difference between 3 and 4 is subtle, but it can be brought out by thinking about the duck-rabbit. Imagine that two people look at the diagram on separate occasions. One reports seeing a duck, while the other reports seeing a rabbit. What exactly should we say about this case? One option is to say that the two subjects *saw exactly the same thing* but interpreted what they saw in different ways. One subject interpreted the drawing as a duck, while the other interpreted it as a rabbit. They saw the same thing, but applied different concepts, we might say. But one could also say that the two subjects *saw very different things*. After all, if you ask the first subject, "What did you see?" she'll say, "I saw a picture of a duck," and likewise the second subject will report seeing a picture of a rabbit. It is very difficult to know just which of these two ways of thinking about this sort of case is the best. Both alternatives have strange unpalatable consequences. For example, consider the apparently weaker claim that the two subjects saw exactly the same thing but interpreted it differently. If that is correct, then strictly speaking, both subjects are wrong about what they saw. The first subject didn't really see a picture of a duck – at least not strictly speaking. She just saw a line diagram that she interpreted as a picture of a duck.

I won't attempt to resolve these issues here. For now, let's just bear in mind that the theory-ladenness of evidence involves claims 1 and 2 above, plus either claim 3 or claim 4, depending on what you ultimately decide to say about the duck-rabbit and similar cases. In *The Structure of Scientific Revolutions*, Kuhn does at least hint that he accepts the stronger claim 4. He describes a fascinating experiment in which some psychologists doctored an ordinary deck of playing cards, adding a red six of spades and a black four of hearts. They then drew cards and asked the subjects to identify what they saw. On different runs of the experiment, they varied the amount of time that the subjects were given to look at the cards. On the shorter exposures, almost everyone identified the anomalous cards as normal. For example, when shown a red six of spades, a subject might simply report seeing a red six of hearts, as if nothing were out of the ordinary. So we might ask: Did the subject see a red six of spades but interpret it as a six of hearts? Or did the subject actually see a red six of hearts? Well, the subject says she saw a six of hearts. And Kuhn writes, rather enigmatically, that "One would not even like to say that the subjects had seen something different from what they identified" (1962/1996, p. 63). Regardless of whether we go with claim 3 or 4, it's clear that the subjects' expectations in this experiment were influencing their reports of what they saw.

Now we can begin to understand what Eldredge and Gould mean when they say that "the data of paleontology cannot decide which picture is more adequate." The problem is that the data are theory-laden. A phyletic gradualist and a fan of PE might look at exactly the same series of fossils and either (3) see the same thing but interpret it in very different ways, or else even (4) see different things. For the sake of argument, let's go with the weaker claim (3). The observational data cannot support one theory or the other unless we know how to interpret the data we're looking at. But we need those very theories to tell us how to interpret the data. A problem of circular reasoning looms. How can the fossil record support PE if we are already interpreting the fossil record in terms of that very theory? It looks like we are presupposing the theory we want to confirm. As Kuhn succinctly puts it, "each group uses its own paradigm to argue in that paradigm's defense" (1962/1996, p. 94).

Indeed, Eldredge and Gould's pessimistic view that the fossil record cannot discriminate between phyletic gradualism and PE echoes Thomas Kuhn's notorious claim that empirical evidence can never serve as the basis for a rational choice between rival scientific paradigms. The theory-ladenness of evidence (or paradigm-relativity of evidence) is just one part of Kuhn's larger case for that quite radical philosophical view. I say "radical," because Kuhn was explicitly challenging the idea that natural science is fundamentally a rational enterprise. In his view, the shift from one paradigm to another is never fully justified by reasons, evidence, and argument; instead, it is more like a religious conversion or an abrupt *Gestalt* shift '– a sudden change to a completely new way of doing things and a new way of seeing the world. Moreover, if you want to know why the scientific community as a group ever abandons one paradigm in favor of another, the best causal explanation is going to appeal to historical and sociological considerations, rather than evidential considerations. In this respect, Kuhn's work wedged the door open for subsequent sociologists and anthropologists of science.

From the very beginning, the Kuhnian influence may have inspired Gould, in particular, to make claims in favor of PE that many critics would find a bit exaggerated or even grandiose. It is easy to see how they – especially Gould – might have thought that by introducing a new way of seeing the fossil record, they had fired the first shot of a Kuhnian scientific revolution. I hope to have shown here that even if Gould did get carried away at times, he and Eldredge really did have an interesting Kuhnian point to make about the fossil record. The relationship between philosophy and natural science is a two-way street.

If PE was inspired by Kuhn's philosophy of science, that does not make it any less valuable scientifically. The fossil record really is a bit like a duck-rabbit. But this raises more questions than it answers. First, how, if at all, can we test the PE model? And if testing paleontology's flagship theoretical innovation turns out not to be possible, what does that say about paleontology? Are there any other interesting theoretical directions in which PE might lead? Are there any ways in which PE might challenge us to move beyond minimalist models of evolution?

3   Punctuated equilibrium

Provocations and problems

> "The data of paleontology cannot decide which picture is more accurate"
>
> (Eldredge and Gould 1972, p. 99)

> "The model of PE is eminently testable"
>
> (Gould and Eldredge 1977, p. 120)

> "The model of PE is scarcely a revolutionary proposal"
>
> (Gould and Eldredge 1977, p. 117)

> "If PE has provoked a shift in paradigms for macroevolutionary theory . . . , the main insight for revision holds that all substantial evolutionary change must be reconceived as higher-level sorting"
>
> (Gould and Eldredge 1993, p. 224)

Gould and Eldredge have said some apparently conflicting things, both about the testability and about the theoretical significance of PE. My aim in this chapter is to get to the bottom of these issues, and to see if there might be some way of making sense of these apparently conflicting claims. I begin with the question of whether PE is all that interesting a model. Then in the second half of the chapter I will move on to consider questions about testability.

PE and macro/micro reductionism

A minimalist model of evolution is one that treats macroevolutionary patterns and trends as mere by-products of microevolutionary processes plus speciation events. Because such models are simplest and most parsimonious, it seems reasonable to start with a minimalist model and expand it only on an as-needed basis. The question is whether any such expansions are needed. In the previous chapter we saw that PE, as originally formulated by Eldredge and Gould (1972), did not require any departures from minimalism. Instead, the initial claim was that different minimalist models of evolution (especially different models of speciation) yield different expectations concerning patterns in the fossil record.

Over the years, Gould, sometimes in collaboration with Eldredge, offered a variety of restatements of PE, responses to critics, and retrospective assessments of its significance. Occasionally, Gould got himself into trouble by making claims that struck many other scientists as exaggerated. Here is one notorious example:

> I well remember how the synthetic theory beguiled me with its unifying power when I was a graduate student in the mid 1960s. Since then I have been watching it slowly unravel as a universal description of evolution. The molecular assault came first, followed quickly by renewed attention to unorthodox theories of speciation and by challenges at the level of macroevolution itself. I have been reluctant to admit it – since beguiling is often forever – but if Mayr's characterization of the synthetic theory is accurate, then that theory, as a general proposition, is effectively dead, despite its persistence as textbook orthodoxy. (Gould 1980a, p. 120)

Passages such as this one have led to a great deal of confusion that is easily cleared up with some close reading and philosophical analysis. It certainly looks as if Gould is claiming here that the neo-Darwinian modern synthesis has been slain by attackers coming from two directions: paleontology, together with the neutral theory of molecular evolution. That earlier development, due to the geneticist Motoo Kimura (1983), suggested that most mutations that occur in the genome are selectively neutral, which means that random genetic drift has a much bigger effect on evolutionary processes than previously suspected. Gould's language invites us to interpret him as proclaiming the death of the modern synthesis. But this interpretation is also highly uncharitable for the simple reason that the synthesis is not dead. Today, in 2010, most evolutionary biologists still work happily within the framework of synthetic evolutionary theory.

It is easy to miss the fact that in the passage quoted above, Gould only makes a conditional (that is, an "if...then") claim. He says only that the synthetic theory is dead provided we work with Ernst Mayr's characterization of the synthetic theory. So without looking at Mayr's characterization, it's impossible to know for sure what Gould is claiming. Gould indeed cites Mayr:

Microeovolution + speciation → macroevolutionary patterns and trends

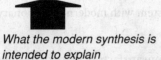

What the modern synthesis is
intended to explain

Figure 3.1 Macroevolution, microevolution, and the modern synthesis.

> The proponents of the synthetic theory maintain that all evolution is due to
> the accumulation of small genetic changes, guided by natural selection, and
> that transspecific evolution is nothing but an extrapolation and
> magnification of the events that take place within populations and species.
> (Mayr 1963, p. 586)

I suspect that Gould misreads Mayr here in one small but important respect.
Mayr is basically saying that the proponents of the synthetic theory (of
microevolution) also accept minimalist models of evolution. At the time that
Mayr was writing, it was probably true that most of the proponents of the neo-
Darwinian synthetic theory also happened to be minimalists. But that does not
mean that minimalism was necessarily part of the synthetic theory. The fact
that all the proponents of some theory *T* accept claim *P*, does not mean that
P is part of *T*. Gould mistakenly – or perhaps only tendentiously – reads Mayr
as saying that minimalism is part of the modern synthesis, whereas in fact
one could easily accept the modern synthetic theory of microevolution while
rejecting minimalism. That is because minimalism is essentially a view about
the relationship between micro- and macroevolution, whereas the modern
synthesis primarily is a theory about how microevolution works (see Figure
3.1). Mayr is just saying that most evolutionary biologists do happen to accept
minimalism. However, if you read Mayr in the way that Gould does, it might
not be quite so crazy to say that the modern synthesis is dead. If you think
that minimalist models of evolution have been shown to be inadequate, and
if you think that minimalism is part of the modern synthesis, then it might
just make sense to proclaim the end of the modern synthesis. The problem
with Gould's pronouncement is that minimalist evolutionary models and the
modern synthesis are two different things.

In an earlier paper, Gould and Eldredge make the following modest, tem-
pered claim on behalf of their theory:

> For all the hubbub it engendered, the model of PE is scarcely a revolutionary
> proposal . . . We merely urged our colleagues to consider seriously the

implications for the fossil record of a theory of speciation upheld by nearly all of us [that is, the allopatric theory], and to recognize the search for phyletic gradualism as a bad historical habit not consistent with modern evolutionary ideas. (Gould and Eldredge 1977, p. 117)

This seems exactly right, but it leaves us in the unsatisfactory place of wondering why PE is so interesting. Gould and Eldredge then follow up this modest characterization of their project with a statement of two consequences. They suggest that if their model holds up, it should lead to changes in the practice of paleontology. Instead of scanning the fossil record for series of specimens that exhibit gradual evolutionary change and ignoring groups of fossils that do not fit gradualist expectations, paleontologists should emphasize "the quantitative study of evolutionary pattern in all members of a fauna" (1977, p. 117). So far, there is nothing much to disagree with. They also claim that PE has one significant theoretical consequence:

"We realized that the extrapolation of PE to macroevolution suggested a new explanation for the fundamental phenomenon of evolutionary trends." (1977, p. 117)

This suggests that, by 1977, Gould and Eldredge had begun to see their model of PE as posing a serious challenge to minimalism. Yet I argued in Chapter 2 that the model, as originally formulated, not only poses no serious challenge to minimalism but even seems to presuppose minimalism. What are the reasons for thinking that minimalism is mistaken? How might PE cause trouble for minimalism after all? How could PE give rise to a challenge to minimalism when Eldredge and Gould's (1972) original argumentative strategy presupposed minimalism?

Jumpy speciation

Gould (1980a, pp. 122–124) made a move that his critics (e.g., Dawkins 1986; Dennett 1995) punished severely and would not let him live down. Yet the move may not have been quite as irrational as the critics would have us think. Gould and Eldredge had originally presented PE as if it were merely a consequence of scaling up the widely accepted allopatric theory of speciation. Now, all of a sudden, Gould begins expressing doubts about the allopatric theory. He points out that the allopatric theory is thoroughly Darwinian, in the sense that it sees speciation as involving the basic microevolutionary processes described by the modern synthesis. In the allopatric model, natural selection has a role to play in causing the geographically isolated sub-population to evolve "away" from the parent population from which it split off. And the evolutionary changes involved are just "cumulative and sequential" changes in gene frequencies. Gould writes that allopatric speciation "is, if you will, Darwinism a little faster" (1980a, p. 122).

By 1980, Gould had begun to wonder whether speciation always occurs in this Darwinian fashion:

> I have no doubt that many species originate in this way; but it now appears that many, perhaps most, do not. The new models stand at variance with the synthetic proposition that speciation is an extension of microevolution within local populations. Some of the new models call upon genetic variation of a different kind. (1980a, p. 122)

This is heretical stuff. Gould is in effect challenging the following orthodox view:

> Microevolutionary change + [further conditions specified by either the sympatric or the allopatric model] → speciation

It's hard to see how one could reject this traditional view without rejecting Darwin's basic idea that cumulative natural selection gives rise to new species. At the very least, Gould is flirting with the idea that speciation sometimes occurs by some other mechanism that's quite different from anything that Darwin described. He is also thinking of the view expressed by the above formula as another kind of "extrapolationism," akin to the extrapolationist (a.k.a. minimalist, reductionist) view of the relationship between micro- and macroevolution.

What other models of speciation could Gould have in mind? Gould (1977; 1980a) writes favorably about the work of mid twentieth-century biologist, Richard Goldschmidt (1940), who argued that speciation sometimes involves an evolutionary jump, leap, or saltation from one species to another, as opposed to a gradual change (at least, gradual on ecological timescales). How might such an evolutionary jump occur? Once in a very long while a new individual shows up in a population, and that new individual has a morphology that differs radically from its conspecifics. This "hopeful monster" will probably die young, but it just might live on to found a new species. If that were to happen, speciation would have occurred not gradually, but in the course of a single generation. Inspired by Goldschmidt, Gould is, in effect, proposing a third model of speciation, in addition to the allopatric and sympatric models: saltational, or jumpy, speciation.

Many critics have dismissed this proposal as crazy and heretical, although it has a few defenders (Thiessen 2006; 2009). Although I don't want to defend it, I do think that it is important to try to get inside Gould's mind, and to appreciate why such a departure from neo-Darwinism might not have seemed so irrational to him at the time. Once you begin looking at the fossil record through the lens of PE, you begin to think that most of the action in evolutionary history takes place during speciation. Once species become established, subsequent evolutionary changes do not typically involve any cumulative directional trends (or so Gould was convinced). Speciation usually occurs quite rapidly in geological terms – though of course that could mean that it still takes tens of thousands of years of gradual evolution. So how does speciation actually work? Well, the allopatric model offers one potential explanation of (geologically) rapid speciation. But neither is it the only model that can explain the geological rapidity of speciation. If speciation were jumpy, or saltational, that could also explain the patterns showing up in the fossil record. This fact alone means that jumpy speciation is a possibility that ought to be investigated, rather than laughed out of town. It's important to think about what it would mean if such an investigation were to pan out (which it didn't). Paleontology, which had previously been on the sidelines of evolutionary biology, would be responsible for a radical rethinking of our understanding of the speciation process.

One problem, as many critics pointed out, is that this jumpy speciation model seems to clash with the standard understanding of how mutation works to give rise to new variation within a population. In the standard picture, mutations involve small, localized changes to an organism's DNA. For example, a point mutation occurs when one nucleotide (a bit of adenine, cytosine, thymine, or guanine) is replaced by another. Often this makes no difference whatsoever to an organism's phenotype. If the point mutation occurs in a non-coding region of the organism's genome, it will have no visible effects. And sometimes, even if it occurs in a coding region, the mutation won't make any difference to the process of protein construction, or to the larger process of building the organism. Scientists often say that mutations are random, but by that they do not mean that mutations are uncaused. All that is meant by "random" in this context is that the mutations are unbiased or undirected. Or to be more precise, the probability that a mutation will occur is unaffected by whether that mutation would help or harm the organism. According to the standard picture, the vast majority of mutations that occur are either neutral or deleterious. Mutations that would actually help an organism in its environment are quite rare. All or nearly all mutations that occur are *micro*mutations (of which point mutations are just one example). Goldschmidt's jumpy speciation model seems to require the occurrence of *macro*mutations: major changes in chromosomal structure that cause equally significant changes in the organism's overall body plan or other phenotypic traits. One challenge, then, is to explain how macromutations could occur. One major difficulty confronting this jumpy speciation model is that we just don't observe this kind of speciation happening in nature. By contrast, the microevolutionary processes involved in the allopatric speciation model are processes that scientists can and do study in nature all the time. Another difficulty is that if we take a given species as our starting point, there will be a vast, huge "space" of possible macromutations that could occur in the genomes of the individual members of that species. Virtually all of those will be very bad news. Intuitively, if you begin with the genome of an organism that is pretty well adapted to its environment and start making changes, the bigger the changes you make, the more likely it is that you will end up with an organism that is not viable at all.

One reason why many biologists reacted so negatively to this proposed jumpy speciation model is that it comes a little too close for comfort to some traditional creationist objections to the very idea of evolution. And indeed, some creationists in the early 1980s unfairly and misleadingly tried to appropriate some of Gould's views. Many creationists who believe that God has created each species in a distinct creative act will grant that microevolutionary processes occur within species. What they deny is that those gradual, microevolutionary changes ever lead to new species. For that, the creationist will argue, you need divine intervention. Now, Gould is not saying that any new species ever appear as a result of divine creative acts. Nor is he denying that ordinary microevolutionary change could ever produce new species – he still has no qualms with allopatric speciation. So even in his most radical anti-Darwinian moods he is nowhere near defending creationism. Nevertheless, Gould was still entertaining the possibility (if not defending the view) that *lots* of speciation is *not* simply a by-product of microevolutionary change over many generations; that is, lots of speciation is not a result of the kinds of microevolutionary processes that we can observe in nature. And that is something that many creationists would agree with.

This comparison with Darwin deniers makes Gould seem extremely heretical. But at the same time, there is another way to spin his flirtation with jumpy speciation. Orthodox neo-Darwinian evolutionary biology recognizes mutation as one of several possible causes of evolution, alongside natural selection, migration, and drift. Macromutation, if it ever occurs, would still be a form of mutation. Thus, in asking us to consider the possibility of jumpy speciation, Gould might be interpreted as making a more modest move: he is proposing to downplay the importance of selection in explaining how new species arise while playing up the role of mutation. This could be understood as a shift of emphasis – though quite a dramatic one – within neo-Darwinian theory, rather than a complete rejection of it. At any rate, Gould was nothing if not a systematic thinker, and it's worth pointing out that the jumpy speciation model is perfectly consistent with two strands of his thinking about evolution that we will encounter again in later chapters: first, his repeated insistence that natural selection isn't quite as important as most Darwinists think, and second, his repeated emphasis on the importance of chance as a factor in evolutionary history.

During the 1990s, a famously acrimonious exchange took place between Gould and one of his most strident critics, Daniel Dennett. Dennett had lambasted PE in his 1995 book, *Darwin's Dangerous Idea*, and Gould responded in kind. I don't want to get too deep into the details of that debate, but arguably one mistake on Dennett's part was to run together the theory of PE with the jumpy speciation model. I think it's important to resist the temptation to conflate the two. The jumpy speciation model does not have much going for it (Dennett was right about that much). But PE, the core of which I sketched in Chapter 2, does not require jumpy speciation. Most scientists today would not accept the un-Darwinian jumpy speciation model, but many do take PE very seriously. We need to understand why. To return now to the earlier theme: the real question is whether PE contains any latent challenges to macro/micro reductionism.

PE and evolutionary trends

PE challenges macro/micro reductionism indirectly by suggesting that some directional evolutionary trends cannot be fully explained in terms of microevolutionary processes. Remember the basic claims of PE from Chapter 2: first,

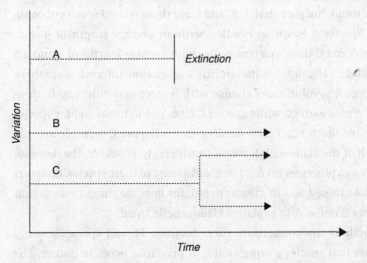

Figure 3.2 Species sorting. A, B, and C differ with respect to some trait. When A becomes extinct and C speciates, the mean value decreases.

species don't change much once they are established. Second, speciation happens relatively quickly (in geological terms). Third, most evolutionary change also occurs rapidly (again, in geological terms) during episodes of speciation. Let's consider Eldredge and Gould's claim about stasis a little more carefully. Eldredge and Gould would never want to claim that species do not evolve at all once they come into existence. Their claim about stasis, rather, is just that the lineage was not evolving in any particular direction between its origination and its extinction. The evolutionary changes that did take place were not cumulative, directional changes.

Suppose for a moment that Eldredge and Gould's main empirical claim is correct (we still have to investigate that) and that most species appear in the fossil record rather abruptly, persist with little or no cumulative morphological change, and then disappear just as abruptly. If stasis is typical in evolutionary history, that means that any large-scale evolutionary trends that do show up in the fossil record cannot be explained as mere by-products of micro-level trends. The reason for that is simply that there typically aren't any micro-level trends. If we want to explain the existence of macro-level trends in the fossil record, we need to look for mechanisms operating at the macro-level. Or in other words, we need to go beyond a minimalist model of evolution.

One kind of macroevolutionary mechanism that might do the trick is *species sorting*, or the differential persistence, speciation, and extinction of entire lineages. Figure 3.2 shows how species sorting can generate a directional evolutionary trend. Imagine that A, B, and C are three related species of coniferous trees. Species A has long needles, with an average length of 4 mm, while species B and C have shorter needles with average lengths of 3 and 2.5 mm, respectively. Let's suppose that during a given time interval, these three species undergo no evolutionary change with respect to needle length. However, species A goes extinct, while species C gives rise to two daughter species which also have shorter (2.5 mm) needles. If this happens, then the average needle length of the clade will decrease considerably. However, the decrease will not be due to processes taking place within any of these species. Chapters 4 and 5 explore these ideas in greater detail; for now the thing to see is that species sorting is one way to explain a cladogenetic trend.

What exactly is the connection, then, between PE and species sorting? No one denies that species sorting could, in principle, occur in nature. The question is whether it ever does occur in nature, and if so, how important a force it has been in evolutionary history. PE suggests that species sorting must have been a significant force, and that it might even have been more significant in some respects than the mechanisms operating at the micro-level. In 1993, Gould and Eldredge published a retrospective on PE in which they seemed to argue that this was the main upshot of their theory:

> The main point may be summarized as follows. Most macroevolution must be rendered by asking what kinds of species within a clade did better than others (speciated more frequently, survived longer), or what biases in direction of speciation prevailed among species within a clade. Such questions enjoin a very different programme of research from the traditional "how did natural selection within a lineage build substantial adaptation during long stretches of time?" The new questions require a direct study of species and their differential success . . . Darwin's location of causality in organisms must be superseded by a hierarchical model of selection, with simultaneous and important action at genic, organismic, and taxic levels. (1993, p. 224)

According to this new hierarchical picture, evolution involves causal processes that operate at different levels. Gould and Eldredge are claiming that Darwin's original evolutionary theory describes the causal processes that operate at just one level, the organismic level. However, if we want to understand the larger-scale patterns and trends of evolutionary history, focusing on that one level will not suffice.

This move toward a hierarchical understanding of how evolution works is arguably paleontology's main contribution to evolutionary theory. Gould, who fiercely advocated for PE through the 1970s and 1980s, was right at the center of this transition, but he was by no means the only important player. Scientists such as Niles Eldredge, Elisabeth Vrba, Steven Stanley, David Raup, and Jack Sepkoski also made major contributions, some of which we will examine in more detail. For now I just want to stress two crucial points: first, if the hierarchical picture of evolution is correct, then minimalist models must go by the wayside. And second, if the hierarchical picture is correct, then paleontology will have contributed to a major revision, or perhaps an expansion, of evolutionary theory.

A tautology problem in paleontology

When Eldredge and Gould first introduced PE, they despaired of ever being able to test their model using data from the fossil record. The problem, as they saw it, was that someone who insisted upon looking at the fossils through the lens of phyletic gradualism would see exactly the same data in a completely different way. For the phyletic gradualist, an abrupt break in a fossil series just shows that the record is incomplete. For a punctuated equilibrian, an abrupt break is historical evidence of rapid speciation. Since there is a disagreement about what counts as evidence for what, the disagreement cannot readily be settled by the evidence. That explains why Eldredge and Gould originally opted to portray their model as a consequence of views about speciation that most scientists already held. Both Gould and Eldredge had spent a great deal of time studying particular fossil groups: Eldredge was an expert on trilobites, while Gould had focused on snails. In their 1972 paper, they showed how the records of these two groups could be interpreted from the perspective of PE. But they also seemed to admit freely that there are other ways of interpreting the same data. Nevertheless, in spite of these early expressions of doubt about testability, paleontologists immediately set out to test the new model. Moreover, Gould and Eldredge subsequently changed their tune. For example, in their retrospective 1993 paper, they laid out several ways of testing PE and suggested that although the issue was not yet entirely settled, the evidence was somewhat in favor of their model.

Some early critics of PE also complained that the model might not be testable, but for a different reason. Levinton and Simon (1980) argued that a central claim of the theory is tautological, and hence not subject to empirical confirmation or disconfirmation. To say that something is tautological is, roughly, to say that it is true by definition. For example, the following propositions all look like tautologies:

All squares have four sides.

Either it's raining outside today or it's not.

Every vertebrate has a backbone.

The meeting at 4 p.m. is earlier than the meeting at 5 p.m.

And so on. Many philosophers have thought that these sorts of propositions don't actually say anything about the world. If you understand them at all, you can see right away that they are true, without having to collect any evidence. A word of caution: I am in fact running together two different kinds of propositions. Some of these are what philosophers would call *analytic* propositions; they are true in virtue of the definitions of the terms they contain. The second one on the list is a bit different. It is true in virtue of its logical form. Statements of the form "P or not P" are always true, no matter what one plugs in for P. Such statements are tautologies in the narrower logical sense. Here I am using the term "tautology" in a broad way to include both analytic truths and logical truths.

The eminent philosopher of science Karl Popper once argued that Darwin's theory is not a genuine scientific theory because there is a tautology at its core (1979; 1996). Popper characterized Darwinism as a "metaphysical research program," rather than a scientific one. Although Popper himself was an enthusiastic evolutionist, creationists everywhere took great joy in pointing out that one of the leading philosophers of science of the twentieth century was saying the same thing about evolution that the scientific establishment was saying about creationism: namely, that it isn't really science. Popper was worried about the following circle of reasoning. Suppose we ask which organisms in a population survive? A natural Darwinian-sounding answer is: the fittest ones. But suppose we now ask which organisms are the fittest. Another Darwinian-sounding answer is: the ones that actually survive! If you define "the fittest" as the organisms that actually survive, then you relinquish the ability to explain why these organisms survived by pointing out that they are fittest. It is just true by definition that they are the fittest. This came to be known as "the tautology problem." There really is a problem here, but only if you define fitness in terms of *actual* survival and reproductive success. It didn't take philosophers of biology long to point out that one need not do that; one can easily define fitness in terms of *expected* survival and reproductive success in a given environment (for helpful discussion, see Sober 2000, pp. 70–73). This is known as the propensity interpretation of fitness, and it has become quite standard.

Levinton and Simon (1980) voice a concern about PE that echoes Popper's concern about evolution. One of the core claims of PE is that once species come into existence, they do not change much until they ultimately go extinct and disappear from the fossil record. But what exactly do we mean by "species" in this context? How do we distinguish between species in the fossil record?

The so-called "species problem" is another major problem area in the philosophy of biology. It turns out that biologists have a number of different ways of thinking about what makes a group of individual organisms into a biological species – or a number of different *species concepts*. In evolutionary biology today, one of the standard ways of thinking about species is due to Ernst Mayr, to whom we also owe the allopatric speciation model. Mayr's idea is widely known as the *biological species concept*, though that is tendentious, since all species concepts are biological. He proposed to count organisms as members of different species just in case they belong to populations that are reproductively isolated from each other. Two populations are reproductively isolated when members of one cannot mate with members of the other and produce viable offspring. This approach to defining "species" is especially attractive because it fits so well evolutionary biologists' understanding of how the speciation process works. Cladogenetic speciation occurs when two populations of a parent species become reproductively isolated from one another. Mayr's view of species is not without its problems. For example, it only applies to organisms that reproduce sexually. And sexually reproducing organisms only represent a fraction of the biological diversity on Earth; much of that diversity is microbial. Another potential problem is that Mayr's concept is next to useless in paleontology. Paleontologists usually have no way of knowing whether two individual organisms whose fossilized remains they are studying belonged to reproductively isolated populations. Suppose, for instance, that they find two very similar skulls which seem to belong to the same type of dinosaur. But suppose that the two skulls were found at sites on different continents in rocks of about the same age. How can anyone be sure whether the animals belonged to reproductively isolated populations? The problem is only exacerbated by the fact that paleontologists often have to make judgments about whether two individuals that lived at different times belonged to the same species. Traditionally, rather than focusing on reproductive isolation, paleontologists have focused on morphology, and on the information about phenotypic traits that the fossil record preserves. On this approach, whether two individuals belong to the same species becomes a question of whether they have the same or very similar morphologies. In paleontological contexts, it is not unusual to define species in terms of morphological similarity. (We might call this the *morphological species concept*.)

Levinton and Simon (1980, p. 37) point out that if we are working with this morphological species concept, then PE seems to be true by definition. Supposedly, one of the main empirical claims of PE is that species do not undergo much morphological change during their "lifetimes." It's tempting to try to test this stasis claim by going out and surveying the fossil record. Although it would be a huge undertaking, we might go species by species and ask, for each species, whether it underwent significant directional morphological change while it existed. Levinton and Simon worry that as long as we are defining species in terms of morphological similarity, then if we did observe significant morphological changes between some specimen A and some later specimen B, we wouldn't count them as members of the same species at all. The stasis claim is trivially true because of the way in which paleontologists think about species. It makes no sense for paleontologists to group fossil specimens together as belonging to a single species on the basis of morphological similarity and then get excited about the "discovery" that species do not exhibit much morphological change. Levinton and Simon write that:

> attempts to document the presence, or absence, of gradual change in characters . . . are confounded with the species identification process. This is particularly true because the paleontologist recognizes a descendant species solely on the basis of change in morphology . . . In answer to the question, "Can rapid morphological change occur in the absence of speciation?" we reply, "in the fossil record, rapid morphological change *is* speciation." (Levinton and Simon 1980, p. 137)

Levinton and Simon are posing a kind of tautology problem for paleontology. If we use the morphological species concept, then the claim that species in the fossil record typically exhibit stasis is true by definition. And if this central claim of PE is true by definition, that means it is not subject to empirical testing.

One response to Levinton and Simon's objection is to distinguish carefully between two issues: first, there is the question of *what it means* to say that two individuals belong to the same species; then there is another question about what *evidence* scientists use to determine whether two individuals are conspecific. In paleontology, the only evidence that scientists usually have to go on is morphological. Morphology is the only available test of species membership, but that is compatible with a variety of accounts of what species membership consists of. For example, one might think that Mayr's biological species concept gives the right account of what it means to say that two fossil specimens belong to the same species, just as it (arguably) gives the right account for living, sexually reproducing organisms. That is compatible with the observation that in paleontology one has to use morphology as evidence of species membership. Once we pull apart the question of what it means for two individuals to be conspecific from the question of what evidence we have for conspecificity, it turns out that Levinton and Simon's claim that "in the fossil record, rapid morphological change *is* speciation" is mistaken. They should have said that in the fossil record, rapid morphological change is evidence for speciation. The tautology problem arises only insofar as paleontologists claim that species do not undergo much morphological change during their existence, while at the same time using the morphological species concept.

Still, Levinton and Simon could counter that because paleontologists do use morphological similarity as evidence of conspecificity, it follows that, in practice, they will never end up classifying two morphologically distinct fossils from earlier and later periods as belonging to the same species. Hence, in practice, they will never end up saying that cumulative morphological change occurs within species. Their methods for grouping fossils as members of the same species seem to rule out the possibility that they will find evidence against the stasis claim. This is not exactly a tautology problem – no longer is there any claim of PE that turns out to be true by definition – but it remains a source of trouble. Levinton and Simon's objection can be reconstrued as the complaint that paleontologists' methods of classifying fossils seem to rule out the possibility of disconfirming the stasis claim of PE.

Another interesting line of response to Levinton and Simon is to describe the respective predictions made by PE and phyletic gradualism without making any reference to species. PE leads us to expect many abrupt, geologically instantaneous morphological changes in the fossil record, as well as long periods of stasis in which morphology changes very little. Phyletic gradualism, on the other hand, leads us to expect mostly gradual, continuous, cumulative morphological change with few if any abrupt breaks or transitions. These predictions about patterns in the fossil record seem readily testable ... or do they? Remember that in Eldredge and Gould's original presentation, both the gradualist and the punctuationist are looking at the pattern described in the lower right, but they are seeing different things. The gradualist just sees evidence that the fossil record is incomplete. But before we conclude that the issue is simply incapable of resolution by empirical study of the fossil record, it's worth pointing out an interesting asymmetry: if paleontologists actually found the patterns that phyletic gradualism predicts, that would seem to count against PE. The easy escape route of appealing to the incompleteness of the fossil record does not seem available to the proponent of PE. After all, if the record did contain lots of series of specimens exhibiting smooth morphological transitions, then the record itself would have to be fairly complete. There would be few missing transitional fossils. This asymmetry makes all the difference; it is the reason why it actually does make sense to go to the fossil record to see what patterns can be found there.

Other criticisms of PE

Another early critic of PE was the respected paleontologist Philip Gingerich, a specialist in fossil mammals. In a 1984 paper, Gingerich tried to raise some problems for the Eldredge/Gould model, and he did so by focusing on early primates, his own area of expertise. He looked at two families of North American primate-like animals from the Paleocene, which lasted from 65–55 million years ago: the carpolestids and the plesiadapids. Both groups are well represented in rocks around Clark's Fork Basin, Wyoming. Gingerich found that new species did seem to show up abruptly, just as Eldredge and Gould had predicted. However, he proposed an alternative explanation: rather than assuming that a rapid speciation event had occurred, why not suppose that animals had migrated in from some other location – say, a location for which we do not have a good fossil record? This first challenge was an interesting new twist on the gradualist move with which Eldredge and Gould were familiar; it is a way of using the incompleteness of the fossil record (in this case, geographical incompleteness) to explain the puzzling patterns we find there.

Second, Gingerich expressed concerns about the mechanism responsible for evolutionary stasis. If you suppose that environments are changing in subtle ways all the time, and hence that selection pressures are also changing, it is puzzling why species should typically remain in existence for so long without exhibiting any directional evolutionary change. If species are not changing much during their lifetimes, why not? Some mechanism seems needed to prevent evolutionary change.

> Lineages evolving according to the Eldredge-Gould model are controlled by some unspecified mystical "homeostatic mechanism" – natural selection in alien environments may have a role to play in overcoming homeostasis during genetic revolutions, but according to the Eldredge-Gould model, it has nothing to do with most evolution. (Gingerich 1984, pp. 126–127)

"Mystical" is a fighting word in scientific contexts, and it may be an unfair term to use here. Gingerich does have a point, insofar as the punctuationist will ultimately need to say something non-mystical about why species don't change much once they come into existence. The basic idea of genetic homeostasis – the idea that Gingerich finds murky – is that "most genes are tied together into balanced complexes that resist change" (Mayr 1988, p. 424). An organism's genes constitute a well-adapted system, an ordered whole; it's difficult to make changes to any one part of that system without messing up the balance (compare also Mayr 1992). Yet in spite of his apparent enthusiasm for the idea, even Mayr frankly admits that he does not know the mechanisms which cause the unity of the genotype (1988, p. 427). We need to be very careful here, however. For one could think that the main claim of PE is a claim about pattern – namely that stasis is typical. One could always remain neutral about the mechanisms that generate that pattern. Another complication is that even though stasis may be the norm, there are some well-documented cases of gradual morphological evolution. So whatever mechanism is invoked to explain stasis – whether genetic homeostasis or something else – will have to be the sort of thing that fails to produce stasis in some cases. And ultimately one would need to offer some account of the conditions that lead to failure.

And yet it does seem worth asking what could explain evolutionary stasis. Notice how this question turns evolutionary theory on its head. Usually we are interested in the causes of evolutionary change; here change is what evolutionary theory leads us to expect, and the absence of change cries out for explanation. One clearly non-mystical possibility is *stabilizing (natural) selection*.

Stabilizing selection occurs when natural selection works against individuals with extreme variations (think of extremely large or extremely small egg size in a population of birds). Maybe birds that lay extremely small eggs (relative to the mean egg size in the population) do less well because their chicks receive fewer nutrients. Perhaps birds that lay extremely large eggs do less well because they have to invest too much energy in producing each egg. If this is the case, then over time natural selection will work to keep the population stable – hovering, as it were, within a stable size range for eggs. One problem for this appeal to stabilizing selection is that PE says that species, once established, typically do not evolve in any cumulative directional way even when environments are changing. It's hard to see why natural selection would not adapt a species to its changing environment. For that reason, it seems like what is needed is a non-selectionist explanation of stasis. Indeed, this is consistent with the spirit of Gould's anti-selectionism. One possibility is that once a species evolves, developmental and other constraints work to prevent further evolutionary change. I'll discuss the issue of constraints at greater length in Chapter 7.

Another possible cause of stasis is gene flow. If a population is large and covers a large territory, then different segments of it will live under slightly different selective regimes. Consider white-tailed deer, which must deal with very different climates in Florida and in New England. One might expect natural selection over long periods of time to drive the different sub-populations of white-tailed deer in different evolutionary directions. But this does not happen. The reason, perhaps, has to do with gene flow. As deer living in one region interbreed with those nearby, genes "flow" from one sub-population to the next, and this has the effect of holding the species together, and preventing any sub-population from evolving in its own direction. Whether gene flow is a powerful enough mechanism to explain stasis, and whether it explains stasis in very many cases, are both open for debate, but it is another example of a non-mystical mechanism that might have some role to play.

Third, Gingerich also reported evidence of gradual cladogenetic speciation in the family plesiadapidae. The rocks he studied in Wyoming contained just what phyletic gradualists would expect to see: clear evidence of a gradual speciation event, with one lineage dividing in two, and the two offspring lineages slowly evolving in different directions. He focused on body size and noted that one of the offspring lineages showed gradual size decrease, while the other showed gradual size increase. This looks like a clear counterinstance to PE – an empirical refutation if ever there was one. Given the asymmetry noted above, it might be tempting to say that PE had been falsified. Indeed, other cases of gradual evolutionary change have been documented, but technically all you need to falsify a general claim is a single counterinstance. If the hypothesis is that all *F*s are *G*s, then you can refute it by finding a single *F* that is not a *G*.

Unfortunately, however (or fortunately, depending on your perspective), Gould and Eldredge never made a universal claim on behalf of PE. They write that:

> We never claimed either that gradualism could not occur in theory, or did not occur in fact. Nature is far too varied and complex for such absolutes; Captain Corcoran's "hardly ever" is the strongest statement that a natural historian can hope to make. Issues like this are decided by relative frequency. (1977, p. 119)

In one sense, they are conceding here that PE is not vulnerable to empirical refutation. Or as a Popperian might put it: the theory seems unfalsifiable. They can just accept Gingerich's example and insist that PE is compatible with there being some cases of gradual speciation. In a deeper sense, though, the theory might still be testable. If paleontologists studied the overall relative frequency of gradual *vs.* punctuated patterns in the fossil record and if the results came out wrong – for example, if gradual change turned out to be the norm, while punctuation and stasis represent the exception – that would be bad empirical news for PE. The problem is that Gingerich, at the time anyway, misunderstood the kind of empirical test that was called for.

This suggests a program of research for paleontologists: they should go out and conduct large-scale statistical studies of patterns in the fossil record. They should try to gain some insight into the relative frequency of gradual *vs.* punctuated evolutionary tempos. What began as a debate between gradualists (like Gingerich) and punctuationists has morphed into something like a Kuhnian normal scientific tradition.

What do the fossils say?

It turns out that the empirical picture is extremely complicated and untidy. Most scientists have agreed all along that the fossil record contains relatively few series of fossils that tell a story of one species splitting into two "off-spring" species via insensible gradations. And most would agree that the fossil record contains lots of examples of rapid morphological change followed by long stretches of morphological stasis. But most would hasten to add that because paleontological studies usually focus on specific groups of organisms and specific time intervals, it is extremely difficult to say anything general about whether evolution is gradual or punctuated. Many studies have generated results that are messy and confusing. Here I'll describe one study that illustrates some of the difficulty in establishing evolutionary tempos.

Malmgren, Berggren, and Lohmann (1983) tried to test the theory of PE by looking at a single evolving lineage of foraminifera, *G. tumida*. Foraminifera are tiny marine organisms (protists, actually) that make up much of the plankton that plays such a crucial role in ocean ecosystems. Foraminifera make tiny shells, often less than 1 mm in length, but the shells fossilize readily, which makes them a good target for paleontological study. Malmgren, Berggren, and Lohmann studied foram fossils obtained from a core sample drilled from the floor of the Indian Ocean. The sequence covered roughly the last 10 million years. To begin with, morphological change in the *G. tumida* lineage was clearly punctuated. The scientists in this case were looking at the size and shape of foram shells. They found that the size and shape remained constant for the first few million years of their sequence, and then changed very rapidly at the Miocene-Pliocene boundary, a little over 5 million years ago. That was followed by another period of stasis. Score one for PE. At the same time, they found no evidence whatsoever of lineage splitting. If any speciation occurred here, it was anagenetic. Indeed, scientists do give different names to the organisms at the beginning and the end of this geological series, based entirely on morphological difference. But without any evidence of lineage splitting, this case study doesn't fit the pattern that Gould and Eldredge tell us to expect. According to their view, rapid morphological change happens during allopatric speciation events, and those events always involve lineage splitting. Malmgren, Berggren, and Lohmann argue that their evidence doesn't fit either the gradualist or the punctuationist pictures very well.

At issue is whether PE represents the "dominant" pattern in the fossil record. The case for that seems to be reasonably good. For my part, I think the best way to keep the empirical issues in perspective is to remember the original problem that led Darwin to complain about the incompleteness of the fossil record: the lack of transitional forms. The excitement that usually accompanies the discovery of such transitional fossils is an indicator of how rarely the fossil record satisfies the expectations of gradualism.

4   Species and macroevolution

"Macroevolution is decoupled from microevolution"

– Steven Stanley (1975, p. 648)

Should scientists go beyond minimalist models to embrace a hierarchical way of thinking about evolution? Are there any large-scale patterns or trends in evolutionary history that are not simply by-products or after-effects of smaller-scale microevolutionary processes? Does evolutionary theory need to include any macro-level mechanisms above and beyond the familiar forces of selection, mutation, and drift?

So far, the case for a hierarchical evolutionary theory might seem pretty solid. To begin with, we have seen that differential extinction and differential speciation can, in theory at least, give rise to cladogenetic trends *even if there is no evolutionary change going on within species at all*. At the same time, the theory of PE suggests that most species, most of the time, do not experience any cumulative directional evolutionary change. If that is correct (and the empirical evidence does not seem too bad), then it points toward the conclusion that most large-scale evolutionary trends are due to differential speciation and extinction – or to *species sorting*. In this chapter, we'll explore this idea in more detail: Can species sorting serve as the basis for developing an interesting macroevolutionary theory? Some paleontologists, especially Stanley (1975; 1979) and Gould, have answered that question in the affirmative. Many other biologists and philosophers are more skeptical. Let's begin by considering just one reason for skepticism.

A reductionist view of species sorting

Microevolutionary theory describes a number of factors – selection, drift, mutation, migration – that can cause changes in gene frequencies in populations. It's quite natural to think that the modern synthesis gives us a clear

picture of the mechanisms of (micro)evolutionary change. Does it make sense
to think of species sorting as a macro-level mechanism in its own right, analogous to those that work at the micro-level? Perhaps not. Suppose we ask what
causes this differential speciation and extinction. One natural answer is that
they are effects or by-products of processes going on at a lower level. Take
speciation, for example: according to the allopatric model, speciation results
from the traditional microevolutionary causes – especially drift and selection –
when those causes operate under special environmental conditions. If something happens to separate a few members of a population from the rest, the
well understood forces of microevolution will do the work of producing a new
species. Similarly, there are perfectly good microevolutionary explanations
for differential extinction. Suppose the climate in a given region becomes
cooler. Species A is an ecological specialist that eats something that only
lives in that region. When the climate cools off, A's food source disappears.
Species B is a generalist; when the climate cools, B is able to exploit different
food sources and so persists. Ultimately, the fitness differences between the
individual members of species A and those of species B explain the differential
extinction. If differential extinction is just a consequence of the fact that all
the As die in the changed environment while the Bs don't, then it seems not
to be an extra mechanism in its own right.

Many paleontologists concede the force of the above reductionist argument.
They tend to think that species sorting, taken all by itself, does not really count
as a distinctive macroevolutionary mechanism. Or to put it another way, even
a staunch minimalist could accept that species sorting occurs, and that some
cladogenetic trends result from differential speciation and extinction. The
minimalist, however, insists that species sorting is itself merely a by-product
of microevolutionary processes. For this reason, those who want to pose a more
serious challenge to minimalism have tended to go in for an even stronger
view. Rather than merely talking about species sorting, they have wanted to
talk about *species selection*. What if there are processes operating at the level of
species that are (almost) perfectly analogous to drift and selection at the level
of individual organisms in a population?

Species sorting, or differential extinction and speciation, could be an interesting evolutionary mechanism even if it turns out to be a mere by-product of
more basic microevolutionary mechanisms. Much depends on the meaning of
"mechanism." According to one widely accepted view, a mechanism consists
of "entities and activities organized such that they are productive of regular

changes from start or setup to finish or termination conditions" (Machamer, Darden, and Craver 2000). Species sorting counts as a mechanism in this sense. It involves entities (species) and activities (speciation and extinction). Macroevolutionary theory tries to show how these entities and activities give rise to regular changes, such as cladogenetic trends. This particular mechanism might not be a basic or fundamental one, but it may still be something that warrants scientific study. It is possible to study species sorting, and to study the ways in which it generates macroevolutionary phenomena, while leaving out, even if only for convenience, all the details about the micro-level.

The MBL model

In the early 1970s, a group of paleontologists gathered at the Marine Biological Laboratory in Woods Hole, Massachusetts. This team, which became known as the MBL group, included Stephen Jay Gould, David Raup, Thomas Schopf, Jack Sepkoski, and Dan Simberloff (an ecologist). Together they set out to transform paleontology into a more rigorous quantitative science, a science that could hold its own when compared with the other mathematical parts of evolutionary biology, such as population genetics. One of the early achievements of the MBL group was the development of a computer simulation of large-scale evolutionary processes, a simulation which is known as the MBL model (Raup et al. 1973; for discussion, see Huss 2009).

At the time when the scientists of the MBL group first met to discuss ways of transforming paleontology into a more rigorous science, computing technology was relatively new. Their first model simulated macroevolutionary branching processes in an extremely simple way. Try to imagine a computer game that proceeds turn by turn. The game opens with just one evolutionary lineage, or species. With each turn, three different things can happen to that lineage:

(1) It can speciate, or branch to form two new lineages;
(2) It can persist to the next round with no change; or
(3) It can go extinct.

In the original MBL model, the scientists stipulated that these three outcomes would be equally probable. The computer used a random number generator to determine what happens to each lineage at each turn. The researchers then conducted experiments – we might call them *virtual experiments* – in which they let the computer simulation run for a large number of turns and waited to see what sorts of patterns emerged. With each run of the simulation, the computer generates a branching tree structure just like Darwin's "tree of life." The MBL group set the program up so that when it first started, the probability of speciation exceeded the probability of extinction. When the number of species in the clade reached a certain point – a kind of pre-programmed equilibrium – the probabilities reset. After that point, the probabilities of speciation, extinction, and persistence become equal. In effect, the computer tosses a three-sided coin to determine what happens to each lineage at each turn.

In later, more sophisticated versions of the model, the MBL group used the computer to simulate the evolution of traits (Raup and Gould 1974). Take, for example, a continuously variable trait such as body size. One can assign a mean body size value to each species and then tell the computer that body size has a certain probability of increasing or decreasing with each time interval.

Like all scientific models, the MBL model involves some idealizations and oversimplifications. Take, for instance, the issue of time intervals. Evolutionary history does not come chopped up into "turns" or "rounds," but the only way to represent it on a computer is to chop it up in just that way. How might the model "map" onto the world? If the model represents actual evolutionary history, then we have to decide, for example, how many years (in real time) correspond to each turn of the simulation. 400,000 years? 500,000? To see just how idealized the model is, imagine that a single turn lasts 1 million years. And suppose that species X goes extinct at turn 17. When *exactly* does species X go extinct? Does it disappear right at the beginning of that million-year period, or sometime in the middle? This question simply has no answer. All anyone can say in this context is that X goes extinct at turn 17.

The MBL model marked a crucial step toward an expanded, hierarchical view of evolution. Notice that the MBL model represents macroevolutionary processes but *not* microevolutionary processes. The designers of the MBL model, especially David Raup, explicitly wanted to see if they could model large-scale evolutionary change while leaving microevolution out of the picture. The initial thought was that if one can study macro-level processes without reference to anything going on at the micro-level, that might suggest that phenomena at the macro-level are not mere by-products of micro-level processes. Notice, for instance, that natural selection – the "force" that many biologists regard as the main cause of evolutionary change – plays no role at all in the MBL model. In a later interview, Raup said that "during the MBL meeting, I guess, the basic question that I must have asked was, what would evolution look like without natural selection?" (Sepkoski and Raup 2009, p. 464). The MBL model simulates the macro without the micro, species sorting without natural selection.

Raup's question of what evolution would look like without natural selection contains an ambiguity. On the one hand, the MBL model does, in a sense, leave natural selection out of the picture. Natural selection simply is not represented in the simulation anywhere. But then the model leaves many things out of the picture – all models do. To give just one more example, the geographical ranges of the species are not represented. On the other hand, there might be an even stronger sense in which the MBL model not only does not represent selection as a cause of evolution but actually assumes that natural selection makes no noticeable causal contribution. This is the difference between (i) not saying anything one way or the other about natural selection, and (ii) saying that selection plays little or no role. Strictly speaking, the MBL model cannot say that selection plays no causal role in evolutionary processes, because in order to do that, the model would have to represent microevolution.

One feature of the MBL simulation that made it so exciting was its *stochasticity*. In the model, whether lineages speciate, go extinct, or persist unchanged is merely a matter of chance. The state of the model at one turn does not uniquely determine the state of the model at the next turn. After the number of lineages in the model hits the specified carrying capacity, what happens to each lineage at each turn is truly random. Extinction, speciation, and persistence become equiprobable. Raup later qualified this in an interesting way:

> The idea was not meant to suggest that things like the extinction of species occur without cause. Rather, that there are so many different causes of extinction operating in any complex ecosystem that ensembles of extinctions may behave *as if* governed by chance alone. (Sepkoski and Raup 2009, p. 463)

The MBL model is what you might call an indeterministic model, in the sense that its earlier states do not determine its later states. This sense of "indeterminism" differs somewhat from traditional philosophical usage (Millstein 2000). Philosophers typically think of indeterminism as the claim that not everything has a cause. Raup is here saying that it is possible to use an indeterministic model (in the scientific sense) to represent evolutionary processes without necessarily embracing philosophical indeterminism.

In the original MBL model, there is a sense in which all species are equally fit. By "equal" here I mean that the probabilities of extinction, speciation, and persistence are the same for all of them. In reality, species differ from one another in two ways. First, the individual members of different species have very different phenotypes and different genomes. Second, in addition to these *individual-level differences*, there might also be *species-level differences*. We might be able to identify variation in features of the species themselves (as opposed to their individual members). For example, one species might occupy a bigger geographical range than another; one might contain more distinct sub-species than another, and so on. The MBL model assumes that these differences between species make no difference to the process of species sorting.

Consider next the analogy between the macro- and the micro-levels. At the micro-level, instead of speciation and extinction, we have the differential reproduction and death of individual organisms in a population. Call this *organism sorting*. Organism sorting will give rise to trends in the population – trends in both gene frequencies and trait frequencies. But why does organism sorting occur? We know that there are lots of (phenotypic) differences among organisms. Do those differences make any difference to the organism sorting process? If so, and if those differences are heritable, then we would say that natural selection is causing the organism sorting. (Or maybe we would say that the organism sorting, in that case, just is natural selection.) If we wanted to, we could easily produce a model that treats all organisms in the population as equal. The assumption would be that the differences among them make no difference to the organism sorting process. If that were the case, then any observed trends in gene frequencies in the population would have to be due to other factors – say, mutation, or more significantly, random drift.

If we take this analogy between the macro- and the micro-levels seriously, it suggests a certain way of thinking about the MBL model. The MBL model is representing species sorting as if the *only* operative factor were drift – or mere chance. So while the MBL model, as I suggested earlier, embodies complete neutrality on the question of whether selection is operating at a lower level – it does not represent microevolutionary processes at all – it does suggest that nothing like natural selection is going on at the level of species. Species sorting is merely a chance process, the analogue of random genetic drift.

The MBL model opens up the door to a distinctive theory of macroevolutionary change that mirrors the modern synthesis theory of microevolutionary change. In the synthetic microevolutionary theory, drift and selection work to produce directional trends in populations. What if macroevolution works in an analogous fashion? What if drift and selection at the species level work to produce trends in clades? If we can make sense of this idea, then we arguably would have a distinct theory of macroevolutionary mechanisms that have been excluded from minimal models of evolution.

Species selection

Steven Stanley (1975) was one of the first scientists to come out strongly in defense of this analogy between micro- and macroevolutionary theory. His influential paper initiated a debate about species selection that continues today. Stanley claimed that species selection "must largely determine the overall course of evolution," and he explained species selection in the following way:

> In this higher-level process species become analogous to individuals, and speciation replaces reproduction. The random aspects of speciation take the place of mutation. Whereas, natural selection operates upon individuals within populations, species selection operates upon species within higher taxa, determining statistical trends. In natural selection types of individuals are favored that tend to (A) survive to reproduction age and (B) exhibit high fecundity. The two comparable traits of species selection are (A) survival for long periods, which increases the chance of speciation, and (B) tendency to speciate at high rates. Extinction, of course, replaces death in the analogy. (1975, p. 648)

Species might well vary in ways that make a difference both to their longevity and their tendency to branch or speciate. Some will have higher probabilities of branching or persistence than others. These probabilities are essentially species-level fitnesses, analogous to the differential probabilities of survival and reproduction of variants within a population.

In the passage just quoted, Stanley draws an analogy between random mutation and the first appearance of a new species within a clade. Referring to Mayr's allopatric theory, he says that there are a couple of senses in which speciation is random. First, speciation usually results from "accidental" environmental changes that lead to the geographical separation of one part of a population from the rest. Second, the isolated sub-population represents a random sample of the genetic variability of the whole population. But these two senses of "randomness" are not the same as the sense in which biologists usually say that mutations are random. To say that a mutation is random is just to say that it does not occur because of the effects that it will have on an organism's fitness. In Stanley's thinking, the species is supposed to be analogous to the individual organism. But no biologist would say that the formation of a new organism counts as a mutation event. And what, at the macro-level, is supposed to be the analogue of the organism's genome? Clearly, the analogy has some soft spots. Nevertheless, Stanley thinks that the randomness of speciation has the following important consequence:

> If most evolutionary change occurs during speciation events and if speciation
> events are largely random, natural selection, long viewed as the process
> guiding evolutionary change, cannot play a significant role in determining
> the overall course of evolution. Macroevolution is decoupled from
> microevolution. (1975, p. 648)

That last statement amounts to a flat-out denial of minimalism, but Stanley strongly overstates the randomness of speciation. Under the allopatric model, natural selection still has a major role to play in adapting the isolated subpopulation to its new environment. One could easily defend the allopatric model without thinking that macroevolution is decoupled from microevolution. Still, many would agree that macroevolution does come uncoupled from microevolution to the extent that we need to invoke species selection to explain large-scale evolutionary trends.

One can easily construct a model, similar to the MBL model, that represents species sorting as resulting from species selection. Consider a simple computer program that starts out with two lineages, A and B. Those lineages get assigned different probabilities of speciation, persistence, and extinction, as in Table 4.1.

The only difference between A and B is that A has a higher propensity to speciate vs. persist with no change. In this model, we can also stipulate that the differential probabilities of speciation vs. persistence are inherited by those species descended from A and B. Here not all species are equal.

Why would A have a higher probability of speciation than B? The speciation and extinction probabilities are determined by facts about the fitness of organisms at the lower level. Kim Sterelny has suggested that a minimalist could well take this line:

Table 4.1 *Species A and B have differential*
fitness because A has a slightly higher probability
of speciation.

	Species A	Species B
Prob (speciation)	.3	.2
Prob (persistence)	.6	.7
Prob (extinction)	.1	.1

> The speciation and extinction probabilities of a species stand in a simple,
> direct relationship to selection on individuals in the populations of which the
> species is composed. For example, if its extinction probability is high, it is
> high because individual organisms are not well adapted by comparison to
> their competitors. (2007, p. 183)

Thus, a minimalist could concede that species sorting really occurs and that
it gives rise to cladogenetic trends. The minimalist could even concede that
different species have different probabilities of extinction and speciation. But
the minimalist can still claim that those differential species-level fitnesses
result entirely from differences among individual organisms. If that is so,
then there is still nothing about species sorting that requires any expansion
beyond minimalist models of evolution. Is there anything else that might
explain differences in speciation and extinction probabilities?

Some of the features of species are mere averages of features of their individual members. Take, for instance, mean body size. We might calculate the
average body size of woolly mammoths and treat that as a property of the
species. It seems plausible that mean body size is a feature that might have
a bearing on a species' probability of extinction and/or speciation. Long ago,
the American paleontologist E.D. Cope speculated that body size might have
something to do with extinction probabilities. Cope was struck by the fact that
at the end of the Cretaceous period, the relatively smaller mammal species
held on, while the bigger dinosaurs went extinct:

> Changes of climate and food consequent on disturbances of the earth's crust
> have rendered existence impossible to many plants and animals, and have
> rendered life precarious to others. Such changes have often been especially
> severe in their effects on species of large size, which required food in large
> quantities. (1974, p. 173)

Cope reasoned that since big animals need much bigger quantities of food,
they would be less able to survive severe changes in their food supply. For

that reason, species with larger mean body size will have a relatively higher probability of extinction. While this may be correct, it poses no real challenge to minimalism. That is because average size is just an *aggregate property*. The mean body size of a species is completely determined by the body sizes of its members. So ultimately, saying that the average body size of a species makes a difference to its probability of extinction is just another way of saying that the body sizes of its members make a difference to the probability of extinction. All you need to do in order to identify an aggregate property at the species level is to pick out one of the traits belonging to individual members of that species and average the value of that trait. Those who believe in species selection as a distinct causal process need something more than this. Species selection seems to require that species have *emergent properties* that are not simply aggregate measures of the properties of their individual members. These emergent properties might make a difference to the probabilities of extinction and speciation, and they might also be passed on to a species' "offspring" via speciation.

If this talk of emergent properties sounds a little mysterious, it might help to consider a couple of examples. One example of an emergent species-level property might be geographical range size (Grantham 2007). Some species have quite large geographical ranges, while others are restricted to very small geographical areas. Geographical range plausibly has something to do with the probability that a species will branch. Another interesting species-level property is *variability* (Lloyd and Gould 1993a). Some species, including our own, are relatively homogeneous, and do not contain much genetic or phenotypic variation. Other species contain a lot of variability. It seems plausible that species with greater variability are more likely to speciate. Could these or other similar emergent properties be the key to understanding species selection? And what exactly is the sense in which these features are emergent, as opposed to aggregate? What exactly does "emergence" mean here? Chapter 5 explores those questions in more detail. First, we should pause to consider a root-and-branch objection to the very idea of species selection.

The individuality of species

Some things in the universe are collections (sets, classes) of other things. Take, for example, my own unimpressive collection of fossils. It consists of a few ammonites and a single piece of stone containing three fossilized fish. This unimpressive collection is just a set containing about half a dozen members.

It is most natural to think of species as sets in this sense; their members are individual organisms (for a defense of this view, see Kitcher 1984). On this model, higher taxa, such as genera, are sets of species.

If a species is a set of organisms, then it seems as if all of the properties of a species would be aggregate properties of the individual members. It might help to think about the properties of my unimpressive fossil collection: it has a certain number of members. The fossils in my collection have an average age, as well as a maximum and minimum age. We could also focus on the dates of acquisition and determine the average length of time that a fossil in my collection has been owned by me. There are countless ways of describing my fossil collection (or any set, for that matter). But it is not easy to think of ways of describing it that do not simply involve aggregating, tallying, or averaging the properties of its members. This line of thought points toward the conclusion that species cannot really be targets of selection. If species are sets, then all of their features are aggregate properties. But if all of their features are aggregate properties, then there are no emergent species-level properties that can make any difference to a species' probability of extinction or speciation.

One response to this objection involves rejecting the idea that species are sets. David Hull and Michael Ghiselin, two philosophers of biology, have long argued that biological species are *metaphysical individuals*. Hull (1988) defines an individual as "any spatiotemporally localized entity which develops continuously through time, exhibits internal cohesiveness at any one time, and is reasonably discrete in both space and time" (p. 26; compare also Ghiselin 1974). According to this view, a species is an individual thing, just like a book, a table, a planet, or an organism. Rather than thinking of individual organisms as members of a species (a set), we should think of organisms as *parts* of the species to which they belong. So you and I, on this way of thinking, are parts of the human species, rather than members of it. Humanity is literally an individual thing that came into existence at a certain point in time in the past, and will go out of existence at a certain future time (hopefully not too soon). It is, in Hull's terminology, "temporally restricted." The human species is also spatially restricted to the surface of one little planet.

The members of a biological species – say, individual human beings – are not spatially contiguous. But most individual things have contiguous parts. Michael Ruse makes this point vividly:

> We think organisms are individuals because the parts are all joined together. Charles Darwin's head was joined to Charles Darwin's trunk. But in the case of species, this is not so. (1992, p. 350)

Yet as Ruse acknowledges, this is not a serious problem. Many individuals have non-contiguous parts. Indeed, physicists tell us that there is a lot of empty space among the atoms that make up our bodies and other individual things. Or to give another example, suppose you remove a key from your laptop keyboard and carry it around all day. The key arguably remains a part of your laptop even though it lacks spatial contiguity with the rest of the machine. The parts of an individual must have certain causal connections to one another, or a certain "internal integration or organization" (Ruse 1992, p. 350). But the individual organisms that make up a species can have the requisite causal-historical connections even if they lack spatial contiguity. At a minimum, each of us is causally-historically related to our parents, for instance.

If species are individuals, then the above objection misses the mark. Intuitively, an individual thing can have features that are not mere aggregates of the features of its parts. Some of the features of an individual thing are aggregate. For example, my weight is just the sum of the weights of all of my parts. But I have other features that are not like that at all. For example, I am the proud owner of half a dozen fossils. None of the parts of me – for example, none of the individual cells of my body – has a fossil collection. Nor is my having a fossil collection an aggregate of any of the features of my parts. Although we have yet to give a precise definition of "emergence," my having a fossil collection looks much more like an emergent than an aggregate property. Thus, if species are individuals, they might well have emergent properties that are relevant to their probabilities of extinction and/or speciation, properties that make a difference to species-level fitness.

At this point, the theory of PE comes back into play. Some have argued that PE lends support to the notion that biological species are individuals. PE implies that species have fairly distinct (geological) moments of origin. They then persist as discrete entities for a certain amount of time before becoming extinct. This makes species look a lot like individual organisms, which also have identifiable starting and termination points. The analogy suggests that species are individuals, too. By contrast, if phyletic gradualism is true and if speciation is sympatric, species will turn out not to have such clear origination or end points in time. Species will fade into one another, and it will not be clear just how to individuate them. So PE could support species selection indirectly by showing that species are metaphysical individuals, as Hull and Ghiselin argue. In this way, the scientists who favored PE and species selection allied themselves with the philosophers who defended the view that species are individuals.

Now the difference between thinking of species as sets and thinking of them as individuals is a matter of *metaphysical categories*: which metaphysical category do species belong to? Unfortunately, Hull conflates two different issues. One issue is whether species are sets *vs.* individuals. A separate issue is whether species (whatever metaphysical category they belong to) are historical *vs.* spatio-temporally unrestricted things. It's easy to conflate the two issues, because individuals typically are historical things with discrete beginning and end points as well as clear spatial boundaries. However, if we felt so inclined, we could also define a set in ways that involve spatio-temporal restrictions. For example, consider the set of all people born on or after January 1, 1697. That set is temporally restricted. Similarly, the set of all people who currently live in Kansas is a spatially restricted set. In light of this, it might be helpful to isolate the question whether species are *historical individuals* or *historically defined sets*.

Does it even matter whether species are individuals or sets, so long as they have distinct temporal boundaries? Think about yourself. On the one hand, you might think of yourself as an historical individual. You came into existence at a given time in the past – maybe at conception, maybe at birth, or maybe somewhere in between. And you will persist through time until you eventually cease to exist. Now perform a *Gestalt* shift, and think of yourself instead as a set of cells. That set is historically defined: its earliest members consist of the cell(s) that constituted you at the time of your inception, while its latest members will consist of the cells you will be made of at the time of your death, whenever that is. In general, whenever it is possible to think of something as an historical individual, it will also be possible, by means of this small *Gestalt* shift, to think of that very same thing as an historically defined set of its parts. Conversely, whenever it is possible to think of something as an historically defined set, it is also possible to think of that set as an historical individual, and to think of the members as parts. The question of which metaphysical category species belong to proves not to be a very interesting one. The difference between historical individuals and historically defined sets is one of those differences that doesn't make any difference.

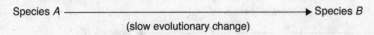

Species *A* ——————————————————————▶ Species *B*
(slow evolutionary change)

Figure 4.1 Gradual anagenetic speciation. Species *A* and *B* lack clear boundaries in time.

Nevertheless, Hull's work does raise another issue that is relevant to the question whether species selection is an important evolutionary mechanism. Suppose we set aside the question whether species are individuals or sets and ask instead: Do species, whatever they are, have *well-defined origination and termination points in time*? Indeed, it is hard to see how species could serve as units of selection if they did not have distinct temporal boundaries. In order to understand this issue, we need to explore three questions: (1) What is the alternative to having well-defined origination and termination points? (2) Do species really have well-defined temporal boundaries? And (3) Why is having well-defined temporal boundaries necessary for being a unit of selection?

Do species have vague boundaries?

The alternative to having well-defined temporal boundaries is having vague boundaries. In order to see what this would involve, consider a case of anagenetic speciation, as depicted in Figure 4.1.

In this case, imagine that species *A* is a variety of bird that lives in an environment where there are lots of seeds and insects. This ancestral species contains a number of different variants, some who specialize a bit more on eating seeds, and some who specialize on insects. Something happens, so that the insects that these birds like to eat disappear. Now selection favors specialization for eating seeds. By the end of this long process, species *B* has become highly specialized. Suppose that other environmental changes occur as well, and that each such change imposes a new selection pressure on the lineage. The descendant species *B* differs considerably from its ancestor. But where exactly does species *A* cease to exist? Where does *A* stop and *B* start? It is difficult to find a non-arbitrary place to draw the line. Anywhere we draw the line, we end up saying that individuals of generation *n* belong to species *A*, whereas the individuals of generation *n* + 1 belong to species *B*. But how can parents and offspring belong to different species? The boundary between the ancestral species *A* and the descendant species *B* is a vague one: the older

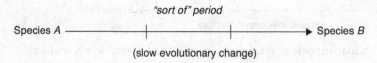

Figure 4.2 Anagenetic speciation and vague species boundaries.

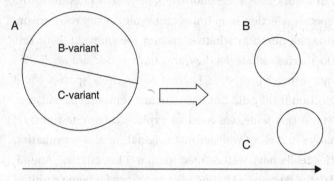

Figure 4.3 Gradual sympatric speciation with vague species boundaries. When does *B* go from being a variant of the ancestral species *A* to being a new species in its own right?

species shades into the new one. There will be a certain historical period that we might call the "sort of" period, as shown in Figure 4.2. If we took a snapshot of the population at this stage, we'd find that the individuals do not obviously belong to species *A* or to species *B*. They "sort of" look like *A*s, but they also "sort of" look like *B*s. Compare: some people are obviously bald. And some people are obviously not bald. But you probably know someone who represents a borderline case – someone who is "sort of" bald. The existence of those in-between cases is what makes baldness an example of what philosophers call a *vague predicate*. The extension of the term "bald," or the set of things to which the term applies, is not well defined. The analogous worry here is that species membership is also not well defined because species have vague boundaries in time.

Denying the reality of anagenetic speciation does not make the problem of vagueness go away. Consider once again the traditional gradualist, sympatric view of speciation. On this view, we start with an ancestral species *A*, which contains two variants, *B* and *C* (see Figure 4.3). Suppose that these two vari-ants make their living in slightly different ways, or that they inhabit slightly different environments within the species' range. Over time, these variants experience different selection pressures, which cause them to differ more and more with each generation. In the beginning, there is just one species with two variants. At the end of this slow process, there are two new species, B and C. At what point shall we say that B and C have undergone enough divergent evolution to count as distinct new species?

Suppose we zero in on the process by which the B-variants gradually evolve into a new species, B. This process will actually look a lot like anagenetic speciation. To begin with, the B-variants might interbreed readily with the other As. But suppose that over time they mix with the other As less and less, and the differences between them and the C-variants become more and more marked. At a certain point we'll want to say that the B-variants have now evolved into a new species – the Bs. But what point is that, exactly? The line-drawing problem we encounter here is the same as the one we ran into earlier with anagenetic speciation. No matter where we draw the line, we'll end up having to say that the members of generation n were B-variants (hence members of species A), while their offspring, in generation $n + 1$, are Bs. This again has more than a whiff of arbitrariness. After all, the members of generation n are just as similar to the members of $n + 1$ as the members of $n + 1$ will be to those in generation $n + 2$. On the gradual, sympatric view of speciation, we'll end up having to say, once again, that there is a "sort of" period – a period of time during which the individuals are "sort of" variants of the ancestral species A, and "sort of" members of a new species, B. Species boundaries remain vague. Talk of a "sort of" period may only make things worse by introducing a problem of higher-order vagueness. When does the lineage go from being B-variants (and members of A) to being B-variants that are also "sort of" Bs?

Darwin himself took the view that species boundaries are vague in just this way. In the following passage from the *Origin of Species*, he seems to say that there is no principled or qualitative difference between a variety, a sub-species, and a species. The difference is simply a matter of degree of evolutionary divergence:

> Certainly no clear line of demarcation has as yet been drawn between species and sub-species – that is, the forms which in the opinion of some naturalists come very near to, but do not quite arrive at the rank of species; or again, between sub-species and well-marked varieties, or between lesser varieties and individual differences. These differences blend into each other in an insensible series. (1859/1964, p. 51)

As I read Darwin here, he is simply stating a consequence of his gradualist, sympatric thinking about speciation. The difference between a species and a "well-marked variety" is vague. This in turn leads Darwin to say things that make it look like he questions the reality of species:

> From these remarks it will be seen that I look at the term species, as one arbitrarily given for the sake of convenience to a set of individuals closely resembling one another, and that it does not essentially differ from the term variety, which is given to less distinct and more fluctuating forms. (1859/1964, p. 52)

I take Darwin's point about arbitrariness here to be a point about vagueness. As in the illustration above, there is simply no non-arbitrary way to look at an evolving group and say, "Alright, up until time t we are dealing with a mere variety, but after that, it's a species."

What about reproductive isolation? Perhaps B and C count as distinct new species, and not merely variants of the original species A, just as soon as they become reproductively isolated from one another. One problem with this proposal is that the notion of reproductive isolation is also vague. Nearly all scientists regard grizzly bears and polar bears as distinct species (*Ursus arctos* and *Ursus maritimus*, respectively). They have different habits, eat different things, and their historical ranges do not overlap much. Yet occasionally, they will interbreed in the wild to produce a hybrid animal known as a "Pizzly" (Roach 2006). These hybrids appear to be quite rare. Hybrids have also been born in captivity, and pizzly bear hybrids in the Lodz zoo, in Poland, turned out to be fertile. So are these two species reproductively isolated? It's tempting to answer: mostly. If in the future their ranges came to overlap more, as might happen if more polar bears are forced onto the mainland by retreating sea ice, it's possible that hybridization could become more frequent. There is a continuum of reproductive isolation (Mallet 2008). At one end, we might place groups that are not isolated at all – say, two segments of a larger population that regularly interbreed. At the other end, we might place groups that could not possibly interbreed and have fertile offspring. Now suppose we ask how much reproductive isolation is enough for two groups to count as distinct species. Although there will be clear cases at either end of the spectrum, there will also be an area in the middle where it's tempting to say that two populations are "sort of" reproductively isolated. If we imagine two variants within an ancestral population diverging very slowly as they evolve, the process of becoming more and more reproductively isolated may also be a gradual one. If so, then even if we use reproductive isolation as a criterion of species distinctiveness, it won't be clear exactly where to draw the line. Older species will shade insensibly into newer ones. For Darwin, this insensible shading, or continuity, was part of the point of evolutionary thinking.

We've now answered question (1). The alternative to thinking that species have well-defined origination and extinction points is the idea that species have vague temporal boundaries, and that there is a non-arbitrary way of distinguishing a variety from a species. But which of these is the correct way to think about species? This leads us to question (2): Do species really have vague boundaries?

The idea that species have vague boundaries is a consequence of Darwinian gradualism together with the sympatric view of speciation. PE, by contrast, seems to imply that species have well-marked beginning and end points in time. Recall that PE takes the fossil record at face value. A central claim of PE is that speciation occurs very rapidly (even nearly instantaneously in geological terms) when a small segment of a larger population rather abruptly finds itself geographically isolated from the rest. The small population size means that the isolated sub-group can diverge quite rapidly and establish itself as a new species in a very short period of time, geologically speaking. Of course, a period of hundreds or thousands of years counts as instantaneous in geological terms. If you could zoom in and watch the process closely as it unfolds, you would still find that it is somewhat vague just when the origination of the new species occurs. Vagueness can never be eliminated completely, but PE does promise to push it into a corner. PE does not imply that there is no vagueness at all; if we zoom in far enough, we'll find that species still have vague temporal boundaries. The point, rather, is that if PE is correct, the vagueness is tightly constrained, and does not show up at all at the level of resolution of the fossil record. As far as the fossil record goes, and as long as we remain concerned with geological timescales, we can say that species have sharp enough temporal boundaries. They abruptly appear and disappear in the fossil record at distinct points in time.

We now need to address question (3). Does this difference between having vague *vs.* sharp temporal boundaries make any difference to the theory of species selection? Consider the following argument:

P1. Species have vague boundaries in time.

P2. Species selection requires that species have distinct boundaries in time.

C. Therefore, species selection is impossible.

Proponents of species selection must block this argument somehow. The theory of PE undermines the first premise. What about the second premise? Species selection and species sorting do seem to require that we be able to say, even if only roughly, where one species ends and the next begins. The reason for that is that sorting, by definition, is a process that involves the coming into existence and the going out of existence of species. If species have vague temporal boundaries – i.e., if ancestral species "fade into" their descendants – then there is no non-arbitrary way of saying when one species has begun to exist and another has ceased to exist. Nor does it help to say that speciation occurs whenever the tree of life "branches," unless we already have a prior understanding of the conditions under which branching occurs. How can anyone say whether a branching (speciation) event has really occurred if we do not know where one species ends and another begins? Although PE does not directly imply species selection, it does help to block one challenging argument against species selection.

5   The case for species selection

Species selection is, as we saw in Chapter 4, one kind of species sorting (compare also Vrba and Gould 1986). Trivially, species sorting is either random (unbiased) or biased. It is just an empirical question whether species sorting is biased or unbiased in any given instance. But it's plausible to think that species sorting is at least sometimes biased. The alternative would be to say that species sorting is always random, and that the MBL model accurately represents macroevolution. Although Schopf (1979) sympathizes with that view, few paleontologists would go so far. Thus, if we define species selection as biased species sorting, it should be relatively easy to document cases of species selection in nature. This is the broadest possible conception of species selection. Some theorists, however, argue that we need a narrower conception of species selection, and that it would be a mistake simply to equate species selection with biased species sorting.

One theme of this chapter is that scientists and philosophers who are interested in species selection confront a trade-off. The more we relax our view of what is required for species selection, the easier it will be to show that species selection really occurs in nature. That means that species selectionists must feel an *empirical pull* toward thinking of species selection in the broadest possible terms. However, they will also experience a *theoretical pull* in the opposite direction. The broader conceptions of species selection hold less theoretical interest because they do not pose much in the way of a challenge to reductionist ways of thinking about evolution. The debate about species selection does not admit of any easy resolution because of the way in which empirical questions (e.g., does species selection really occur?) interweave with conceptual questions (e.g., what exactly does species selection require?). Questions about species selection sit right in that boundary region where philosophy and natural science overlap.

The debate about species selection has unfolded as part of a much larger debate within evolutionary biology that reaches beyond paleontology: the debate concerning *levels of selection*. That larger controversy was instigated in the 1960s and 1970s by a group of reductionist-minded biologists, including most famously George Williams (1966) and Richard Dawkins (1976), who argued that genes are the ultimate targets of natural selection. In Dawkins's memorable way of putting it, organisms are mere "vehicles" or "survival machines" that genes construct in order to help ensure their success in competition with other genes. In the 1970s, when paleontologists such as Stanley, Vrba, and Gould first began to articulate and defend what they sometimes called the "hierarchical expansion" of evolutionary theory, they did so in a hostile intellectual environment. Gene selectionism ruled the day, and many biologists saw it as a tough-minded response to earlier, carelessly formulated ideas about evolution working for the good of the group, or the good of the species. The idea of group selection, which even Darwin had flirted with in his discussion of human evolution, had fallen on hard times. And species selection, in a sense, was even further "out there" than group selection. In the meantime, though, the intellectual climate has changed considerably. Biologists and philosophers of biology began to identify problems with gene selectionism (for a helpful introduction, see Sterelny and Griffiths 1999, Chapters 1–4). What's more, during the 1990s, group selection made an unexpected comeback. Today, most evolutionary biologists remain open to the suggestion that natural selection can operate at different levels of the biological hierarchy – on genes, organisms, and groups – at the same time. Most now treat it as an empirical question just where and how selection is operating in a given case. Okasha (2006) offers an up-to-date exposition of what has come to be known as multi-level selection theory. In the current intellectual climate, species selection no longer seems quite as radical as it once did.

Yet even though the general climate in evolutionary biology today is much friendlier toward species selection than it once was, we will see that some of the fundamental issues remain unsettled.

An argument from PE to species selection

In Chapter 4, I argued that there are a couple of respects in which PE *suggests* species selection, without actually implying that species selection really occurs. For example, PE suggests that species, once in existence, typically do not undergo any persisting directional morphological evolutionary changes. Thus, if there are any cladogenetic trends in evolution, those trends may not be explicable in microevolutionary terms. This is not a clear argument for species selection, in part because the trends could be explicable in terms of species sorting, without any genuine species selection taking place. Cladogenetic trends could also result from the rapid morphological change that, according to PE, takes place in conjunction with speciation.

Yet there is another, closely related line of argument that some of the proponents of PE and species selection made. That argument requires one additional assumption that Gould and Eldredge (1977) referred to as "Wright's rule," named in honor of evolutionary biologist Sewell Wright. Wright's rule says that "speciation is essentially random with respect to the direction of a macroevolutionary trend" (Gould and Eldredge 1977, p. 132). When a new species first arises, the speciation event will involve significant morphological change. For example, the new species could be either larger or smaller than its predecessor. Wright's rule, when applied to this example, says that size increases and size decreases are equally probable. When Gould and Eldredge invoked Wright's rule, they may have had the MBL model in mind. The original MBL group used their simulation to model the evolution of traits as well as phylogenetic branching processes (Raup and Gould 1974). One thing they did was to make increases or decreases in the value of some trait (such as body size) random in order to see what would happen. Assuming that Wright's rule holds true is similar to assuming that species sorting is unbiased; the difference is that Wright's rule concerns the direction of morphological change rather than the probabilities of speciation and extinction. Notice also that Wright's rule closely parallels the received idea that mutations are unbiased, or undirected, or (in a sense) random with respect to natural selection. Eldredge and Gould saw that if Wright's rule were true, it could serve as the basis for an elegant argument for species selection:

P1. There are cladogenetic trends, such as trends in body size. Indeed, such trends are quite common. (This is an unproblematic empirical claim.)

P2. Nearly all the morphological change that occurs in evolution occurs during speciation events. (This is one of the central claims of PE.)

P3. But the direction of the change that occurs during speciation events is random. (Wright's rule.)

C. Species selection must be quite common, because it is the only way to explain cladogenetic trends.

This argument goes beyond the mere suggestion that species selection accounts for some trends. But the argument is only as convincing as premise 3. Why should we think that Wright's rule holds true? Of course, the evolutionary changes that coincide with speciation events *might* be entirely random, in the way that Wright's rule suggests, but there is no clear evidence that they typically are. Okasha (2006, pp. 204–205) points out that species selection could still occur even if Wright's rule turns out to be false. Since Wright's rule is so dubious, it makes more sense to defend species selection in a more straightforward way – namely, by trying to point to examples of it.

It is also worth pointing out that Wright's rule is not an essential component of the theory of PE. PE only says most morphological change occurs in spurts, during speciation events. PE, to repeat a theme from earlier chapters, is mainly a theory about the tempo of evolutionary change. One can endorse PE without saying that the direction of morphological change is random, just as one can endorse PE without saying that species sorting is always random.

These issues are important because they concern the relationship between PE and species selection. Those two theories are often presented as if they were part of a single package that one can take or leave. That may be due to the fact that both are so closely associated with the work of Gould. But if I am correct, they are best seen as freestanding theories that are logically independent of one another. PE does suggest species selection, but that is as far as the connection goes.

Does species selection explain the prevalence of sexual reproduction?

The earliest angiosperms, or flowering plants, appear in the fossil record in the early Cretaceous period. By about 120 million years ago, the flowering plants were well established and had begun to diversify (Friis et al. 2006). The earliest flowering plants show up quite abruptly in the fossil record, and Darwin, who called the origin of angiosperms "an abominable mystery," worried about the absence of intermediate forms between the angiosperms and the gymnosperms which had dominated the earlier Jurassic landscape (Frohlich and Chase 2007). The fossil record does not tell a clear story about how and why flowering plants first evolved, but it does show that flowering plants rapidly evolved and took over during the early to mid Cretaceous, and that they have remained the dominant (at least, the most species rich) group of terrestrial plants ever since. One of the originators of the idea of species selection, Steven Stanley (1979), argued that species selection best explains the prevalence of sexual reproduction in plants and other groups.

Stanley (1979, Chapter 8) uses the theory of PE to generate a novel argument for species selection. He reasons that "the punctuational model holds that rapid evolution is concentrated within speciation events, yet asexual taxa cannot speciate in the normal sense" (1979, p. 215). In order to appreciate this point, it might help to consider the evolution and diversification of two plant species that are placed in new environments. We can suppose that the two species are pretty much alike, except that one reproduces asexually and the other sexually. In the asexual species, offspring are genetic clones of their parents. Advantageous variations get passed on from parents to offspring, but let's suppose that there is no lateral gene flow or recombination of genetic material. Evolution and diversification will be "sluggish," according to Stanley. Diversification will still occur, but it will occur very gradually as different variations accumulate in different sub-groups of the population. Allopatric speciation, as described by Mayr, will never occur. In fact, it is a vexing question just what is meant by "species" in asexual organisms in the first place. But assuming that it does make sense to group them into species, it seems that speciation in a group of asexual plants would have to fit the gradualist picture. Now contrast that with the sexually reproducing plants. At first, the gradual diversification of the asexual plants might even outpace the sexually reproducing ones – or so Stanley argues. That is because (he assumes) species remain in stasis throughout much of their lifespans. But the sexually reproducing species has the potential to diversify rapidly to exploit new environmental resources. Stanley argues that this difference between asexual and sexual reproduction can make a big difference.

The difference between the two groups has to do with speciation rates. In effect, Stanley's argument boils down to this (1979, p. 215):

P1. Species with sexual reproduction will have considerably higher speciation rates than asexual species. This is the insight supplied by PE.

P2. But at the same time, asexual and sexual species will typically have the same extinction rates.

C. Therefore, species selection will generally favor sexual species over asexual ones. In a word, sexual reproduction enhances species-level fitness by increasing the speciation rate.

This means, in turn, that species selection has the potential to explain the prevalence (though not the first appearance) of sexual reproduction.

In order for the argument to go through, Stanley would need to answer some additional questions. First, why should we think that species selection affords the best or the only adequate explanation of the prevalence of sexual reproduction? Is there no good way to explain the prevalence of sex without appeal to species selection? Second, if species selection generally favors sexual reproduction, why do so many asexual species remain? In his original presentation of the argument, Stanley does attempt to address these issues. For example, he argues that traditional gradualist models have trouble explaining the prevalence of sexual reproduction, and that there might be other ways of explaining why asexuality has remained prevalent in some groups, such as prokaryotes. These are complicated issues, and I propose to set them to one side in order to focus on two further problems with Stanley's view. These problems are serious enough to force us to look elsewhere to try to build a case for species selection.

First, there is a conceptual problem at the heart of Stanley's argument. To begin with, it is by no means clear that asexually reproducing organisms come grouped into species in the same way that sexually reproducing ones do. It's not clear, in other words, that Stanley's asexual and sexual species are even the same sorts of things. It is well known, for example, that Mayr's biological species concept, which emphasizes reproductive isolation, does not apply to asexual organisms. Certainly asexual and sexual species don't speciate in the same manner. But that leads to a problem: we can't talk univocally about speciation in asexual *vs.* sexual organisms. Stanley's model requires that we be able to compare speciation rates in both groups, but "speciation" has different meanings in asexual *vs.* sexual groups. To approach the same problem from another direction: recall from Chapter 4 that PE sets the stage for species selection by showing that species have relatively clear boundaries in (geological) time. In order to have any clear sorting process at all, we need to be able to individuate the items being sorted. It is far from clear, however, that asexual species have temporal edges that can be drawn in a non-arbitrary way. It isn't clear, in other words, that asexual species satisfy the necessary conditions for species selection in the first place. Stanley writes that "in effect, sexuality represents a virtual *sine qua non* for success in species selection" (1979, p. 215). It arguably also represents a *sine qua non* for participation in species selection in the first place.

Suppose, however, that we can find a way to make sense of the idea that asexual *vs.* sexual species have different speciation rates. This would amount to a difference in species-level fitness. That's just what is needed for biased species sorting. What explains the differential speciation rates? According to Stanley, the difference is due to sexual reproduction. Strictly speaking, though, sexual reproduction is a feature of individual organisms. We only call a species "sexually reproducing" because its individual members reproduce sexually. This is like calling *Tyrannosaurus rex* a big species; what we really mean is that the individual Tyrannosaurs were big. In this case, it looks like the differential speciation rates are due to differences in the features of the organisms, not to differences in species-level properties. In case this distinction between organism-level properties and species-level properties is unclear, consider an example of a species-level property, such as population size. An individual Tyrannosaur does not (and could not) have a population size. Population size is a clear example of a species-level property. Sexual reproduction is a property of individual organisms; it isn't like population size. Species obviously do not reproduce sexually, by mating with other species and having offspring, but individual organisms do.

The distinction between species-level properties and organism-level properties is a metaphysical one, and for that reason it might seem unrelated to any live scientific issues. However, it makes all the difference to the question whether species selection really goes beyond minimal models of evolution. If the differences in speciation rates (or in species-level fitness) are entirely due to differences in the properties of individual organisms, then the species-level sorting process remains fully explicable in terms of processes that operate at the level of organisms and populations. The macro remains a mere by-product of the micro. Of course, whether this is a real problem for Stanley's argument depends entirely on what one hopes the argument will show. Stanley himself (1975) favored a hierarchical view in which macroevolution is "decoupled" from microevolution. His argument concerning the prevalence of sexual reproduction does not lend much support to that view.

Elisabeth Vrba on "effect macroevolution"

It seems plausible that biased species sorting occurs in nature. The alternative – namely, the view that species sorting is always totally random – seems extremely far-fetched. Is that perhaps all one needs to say in defense of species selection? Unfortunately, things are not so simple. In a series of papers from the 1980s, Elisabeth Vrba showed that biased species sorting can sometimes be a kind of "incidental effect" of natural selection working at the level of individual organisms (Vrba 1983, 1984; Vrba and Eldredge 1984). She called this idea the "effect hypothesis" – a somewhat infelicitous term, since most scientific hypotheses say something about causes and effects. But the basic idea is clear enough: biased species sorting can be a side effect of microevolutionary processes.

Vrba's "effect hypothesis" drew inspiration from her study of the evolution of African antelopes (Vrba 1987). She compared the evolution of impalas with that of wildebeests over the last several million years. One initial reason for looking at antelopes is that it is easy to distinguish different species in the fossil record based on horn morphology. She found that if you count individual organisms, impalas are far more numerous today than are wildebeests. So in that sense, it might seem that impalas are more successful. Over the last 5 or 6 million years, however, the two groups have had very different evolutionary histories. The impalas have barely speciated at all; only one or two impala species existed in the timeframe that Vrba studied. By contrast, she identified several dozen different species of wildebeests. This is as clear an instance of biased species sorting as anyone could ask for. Wildebeests had a much higher speciation rate than impalas. But why?

Vrba suggested the following explanation for the difference: impalas are ecological generalists who can take advantage of different kinds of plant communities. Wildebeests, on the other hand, are much more highly specialized for taking advantage of certain kinds of foods. Natural selection has adapted them for life in a very special kind of open grassland habitat. Vrba showed that this specialization could have an interesting side effect. At times in Africa's recent geological history, climate changes have led to changes in patterns of vegetation, dividing up the wildebeest's preferred grazing habitat into islands. Populations of wildebeests would sometimes find themselves isolated from the rest of the species – just the kind of geographical isolation needed for speciation according to Mayr's allopatric model. The high speciation rate of the wildebeests was just an "incidental effect" of the fact that natural selection had specialized them for eating certain kinds of plants.

In effect, Vrba draws a distinction between biased species sorting and species selection. She points out that biased species sorting can occur as a result of natural selection working at the level of organisms, and she distinguishes this phenomenon – which she calls "effect macroevolution" – from *bona fide* species selection. What else is needed, in addition to biased species sorting, for species selection to occur? One natural suggestion is that the biases in species sorting must be due to species-level traits that are, in some sense, *emergent*. This move has led to what might be called the "emergent character approach" to understanding species selection.

Before moving on to explore the emergent character approach, we should pause to consider Vrba's scientific motivations. She herself is interested in defending the idea of species selection, but she adopts a restrictive view of what species selection involves. She is, if you will, responding to what I earlier called the theoretical pull. At the same time, she argues that effect macroevolution is a highly interesting phenomenon, even though it falls short of qualifying as real species selection. Effect macroevolution has one extremely important consequence for our thinking about larger-scale evolutionary trends (which will be the subject of upcoming chapters). It's most natural, perhaps, to think that when there is a larger-scale, cladogenetic trend in a certain direction in evolutionary space, natural selection must get the credit. The trend must lead in the direction of increased adaptedness to the environment. Vrba's notion of effect macroevolution implies that even though natural selection may ultimately deserve the credit for the larger-scale trend, the larger trend or pattern may only be a kind of accidental side effect of selection. The fact that selection is the cause says nothing about whether the trend involves increasing adaptedness, or any kind of evolutionary improvement.

The emergent character approach

From a theoretical point of view, the most interesting case of species selection would seem to be one that meets the following conditions:

1. Species sorting is biased;
2. The bias is due to differences in species-level traits;

3. Those species-level traits are heritable, in the sense that "parent" species pass them on to their descendants; and

4. The species-level traits that make the difference are *emergent*.

If all of these conditions were met in nature, then that might seem to clinch the case for the hierarchical theory of evolution. Of course, it could still turn out that species selection (in this strong sense) is extremely rare, and hence not too important for understanding the grand sweep of evolutionary history. The discussion of species selection could easily morph into what John Beatty calls a "relative significance debate" (1995; 1997). Perhaps all parties would concede that species selection sometimes occurs; then the points of contention are how often it occurs, and how much it matters. But before we get into such a debate about the relative significance of species selection, proponents of the idea need to point to some clear-cut examples of species selection in action. Stanley, as we just saw, tried to do that, but the case he offered failed to meet the fourth condition above. Sexual reproduction is not an emergent, species-level trait.

I will turn shortly to consider another putative case of species selection explored by paleontologist David Jablonski and philosopher Todd Grantham. First, though, we should pause to consider the meaning of "emergence" in this context. This is a contentious issue, and it is one place where the philosophy of paleontology links up with much bigger issues in the philosophy of science and beyond. Without a workable definition of "emergence," we won't know what condition 4 above really amounts to. Table 5.1 presents several different senses of "emergence" discussed by Grantham (2007).

The philosophical literature on emergence and the related notions of reduction and supervenience is vast, and these four options by no means exhaust the field. The important thing to see is that emergence comes in different strengths. Some senses of "emergence" are too weak to do the work that species selectionists need; other senses are too strong. The trick here, as so often in philosophy, is to produce a theoretical definition of "emergence" that is strong enough to do the necessary work but not so strong that it strains plausibility.

Philosopher of biology William Wimsatt (2007) contrasts emergent properties with aggregate properties. Up to now, I have been using the term "aggregate property" mainly to refer to species-level properties that are mere sums or averages of the properties of the individual members of the species. Wimsatt uses the term in a narrower and more demanding sense. He begins by posing the following question:

Table 5.1 *Varieties of emergence.*

Wimsatt (1997; 2007)	An emergent property is some property of a whole system that depends in some way on the organization of the system.	Too weak
Vrba (1989)	An emergent property is some property of a system which is not the kind of property that the components could have.	
Weak emergence Bedau (1997), Grantham (2007)	A property of a system is emergent when (a) it is emergent in the above two senses, and (b) it is causally incompressible.	Just right?
Strong (metaphysical) emergence	A property of a system is emergent when it can undergo changes without any corresponding changes in the parts.	Too strong, too mysterious

What if some properties of the parts and system were invariant no matter how you cut it up, aggregated, or rearranged its parts? For such properties, organization wouldn't matter. (2007, p. 175)

To illustrate this, Wimsatt asks us to imagine running an organism through a blender. That process would destroy almost all of the organization of the organism, leaving a blended mess. The blending would also destroy most of the organism's properties. But perhaps some properties, such as mass, would remain invariant through this process. Those properties that remain invariant are the purely aggregate ones – I will call them *W-aggregrate* properties (short for "aggregate in Wimsatt's sense"). W-aggregate properties are those features of a system that remain invariant under *any* decomposition, re-arrangement, and/or substitution of its parts. Wimsatt acknowledges that this W-aggregativity is "rare indeed" (2007, p. 280). He then proceeds to define emergence as any failure of W-aggregativity. This means that most of the properties of organisms count as emergent, since few of the organism's properties will survive the blender. Many properties of species will turn out to be emergent as well. Wimsatt readily acknowledges that this is a rather weak conception of emergence; he writes that "emergence in this sense is thus extremely common" (p. 175). But he argues that there is interesting work to be done when it comes to analyzing the different ways in which W-aggregativity can fail.

Grantham (2007) argues that Wimsatt's notion of emergence as the failure of W-aggregativity is too weak to be of much help to species selectionists. Wimsatt himself notes that emergence, in his sense, is compatible with explanatory reduction. He suggests that a reductive explanation of some feature of a whole system "is one that shows it to be mechanistically explicable in terms of the properties and interactions among the parts of the system" (Wimsatt 2007, p. 275). Organisms have many properties that can be reductively explained in this manner, and yet still would not survive the blender. Suppose the organism in question is a plant with the ability to fix nitrogen in the soil. It loses that ability when it goes through the blender. Still, scientists can give a fully adequate mechanistic explanation of the ability of the plant as a whole to fix nitrogen in terms of the properties of its parts as well as the symbiotic bacteria that do the work. On Wimsatt's account, a property such as the ability to fix nitrogen could be reducible and emergent at the same time. Species selectionists who adopt the emergent character approach would seem to want something more than this. They want to try to explain species-level fitness differences in terms of species-level properties that are not reducible to the properties of individual organisms.

Species also have properties that are not even the kinds of properties that individual organisms could have. Take, for example, sex ratio or mean body size. An individual organism might have a sex, but obviously could not have a sex ratio. It could have a body size, but not a mean body size – unless, I suppose, you measured its body size at different times and calculated the average of those measurements. Once again, this sense of emergence is too weak. Properties such as sex ratio and mean body size are pretty obviously aggregate properties (in the ordinary statistical sense of "aggregate"), and they are fully explicable in terms of the properties of individual organisms. Indeed, talking about the sex ratio of a species is arguably just a shorthand way of talking about the sexes of all the individual members.

These first two senses of emergence are too weak because a species-level trait could be emergent in either sense and yet fully reducible to the traits of individual organisms. We should also be wary of going too far in the other direction and defending a sense of emergence that is too strong. Whatever we say about species-level properties, we want them to fit nicely into a broadly physicalist ontology. A physicalist is someone who holds that everything is physical. That is, everything real is composed of the entities, processes, and properties described by fundamental physics. (For an accessible overview, see Post 1991.) One traditional physicalist slogan is: "No difference without a physical difference." Applied to the case of species, that means that there can be no difference in species-level properties without some difference in the properties of individual organisms. The individual organisms, after all, are the physical things that species are made of. One could define an emergent property in a strong way, as a higher-level property that can change without some corresponding change in the lower-level properties. In other words, one could define an emergent property as one that violates this or some other physicalist stricture. But that would be too strong. Presumably, species selectionists have no interest in challenging physicalism.

What's needed, then, is an account of emergence that is "just right": the species selectionist needs to define "emergence" such that emergent species-level traits are not fully explicable in terms of the traits of individual organisms. But the account must not be so strong that it makes the species-level properties non-physical or ontologically mysterious.

Species selection and geographic range

Jablonski (1987) argues that there is evidence of species selection taking place in two groups of marine mollusks from the last 16 million years of the Cretaceous period: gastropods and bivalves. To be more precise, Jablonski found a clear correlation between a species' duration – the amount of time it lasted – and its geographic range. The fossil record for marine invertebrates contains ample information about both of those quantities. Intuitively, it makes sense that longevity and geographic range might be related. Having a larger geographic range might help to protect a species against extinction; if times get tough in one part of the species' range, the species may yet hang on in another geographic region. Jablonski also asked what was causing what. Was geographical range exerting an influence on duration – say, by lowering the probability of extinction – or did the causal relationship work the other way around? Could it be that having a long duration is what enables species to achieve a wide geographical dispersal? To test these ideas, he looked at a number of species that evolved just before the Cretaceous-Tertiary mass extinction event. Those species, like the others, went extinct at the K-T boundary. But they had achieved a large geographical range before going extinct, even though they had not yet been around that long, in evolutionary terms. That suggested to Jablonski that duration is not the cause of wide geographical range, but rather vice versa. This is exactly the kind of case study that species selectionists need.

Indeed, geographical range clearly meets the conditions laid out in the previous section. It is a species (or perhaps population) level property. Strictly speaking, it is not the sort of property that an individual organism can have. We might say that an individual organism has an *ambit*, defined perhaps as the geographical territory that the organism covers in its lifetime. Range is an aggregate spatial concept that we arrive at by combining all the ambits of the individual members of the population. Range is also emergent in Wimsatt's weak sense. The range of the species does not remain invariant no matter what operations you perform on the individual members of that species. Finally, range is a metaphysically respectable notion that conforms to the physicalist's "no difference" principle. It is impossible to vary the geographical range of the species without some change in the ambits or habits of its members. Not only that, but Jablonski showed that range size is a heritable property, meaning that species tend to have range sizes that are similar to the species they evolved from. He showed this by carrying out a heritability analysis on sister species – species that were closely related phylogenetically. Sister species tended to have similar range sizes. It therefore looks as if all the conditions for species selection fall into place in Jablonski's case study. Indeed, Jablonski himself argued that species selection would generally favor the evolution of larger range sizes, because species with larger geographical ranges would for that reason be fitter – i.e., less prone to extinction.

If this were the whole story, then the case of geographical range size might not be one in which macroevolution is uncoupled from microevolution. However, Grantham (2007) takes the argument a step further. Drawing on some earlier work of Mark Bedau (1997), he claims that geographical range size is emergent in another interesting sense because it is "causally incompressible." Grantham explains this notion as follows:

> One crucial point is that [the species-level property P] cannot be predicted from the lower level without simulating the full suite of processes that have historically led to P. In other words, the dynamical processes that determine P are "causally incompressible." (2007, p. 79)

Grantham's idea seems to be something like this: suppose you wanted to predict the geographical range size of a given species at a given time. Since geographical range size is a result of historical processes, you must base your prediction on what you know about the history of the species, and the processes that led to it having the range size that it does. However, imagine that you have no access to any information about this particular species' historical range size. In fact, you have no direct information at all concerning the properties of the species. What you do have access to is all the relevant information about the individual members of that species, including information about their movements, their locations, and their ambits. Those are the lower-level properties that, in some sense, constitute the species-level property of geographical range. Now clearly the range of the species at a given time is a result of all the locations, movements and ambits of all of its individual members up to that point in time. If you had all of that information about historical processes at the individual level, you could predict the species' geographical range. Of course, in order to do that, you would have to have cognitive super-powers or else a powerful computer simulation; we are talking about far too much information for a single human mind to handle. Is there any shortcut way of predicting the geographical range of species while relying only on information about the individual organisms? Is there any way of compressing the information about the spatial properties of the individuals? If the answer is no, then we have another sense in which geographical range can be said to be "emergent." Grantham claims that this sense of emergence is just what species selectionists need, because it is stronger than the senses discussed earlier, and yet not so strong as to involve any departure from metaphysical respectability.

In case Grantham's notion of causal incompressibility is not entirely clear, we should consider a case in which the relevant historical processes are causally compressible – a case to use as a foil for thinking about geographical range. Grantham offers the following example:

> Suppose, for instance, that given full information about the initial conditions and selective forces operating on a large population, we can accurately predict the equilibrium sex ratio of a population without tracing its full causal history. Even if sex ratio has further causal effects, it is not unreasonable to view the causal chain "organismic selection regime → sex ratio → effect" transitively, so that further effects can be explained reductively as a consequence of organismic selection. Essentially, the organismic selection

regime provides a "proxy" for sex ratio (at least within a certain range of conditions) so that we can explain the effects of sex ratio as remote effects of this selection regime. (2007, pp. 79–80)

What if we could predict the sex ratio without knowing all the gory details about the individual-level historical processes? What if all we needed was information about the initial state of the population plus the selective forces acting on it? And what if the sex ratio, in turn, makes a difference to species-level fitness? Grantham's point about the transitivity of causation is that in such a case, the differences in species-level fitness (along with any further effects of the sex ratio) would simply be due to the initial conditions and the selective forces. Grantham argues that where you do have causal compressibility, as in this hypothetical case, there is a sense in which the higher-level effects are reducible to the lower-level causes. He thinks that this is an interesting kind of explanatory reducibility that is not present in cases where the lower-level historical processes are not causally compressible. Conversely, there is an interesting kind of emergence present in cases where there is no causal compressibility.

Grantham's point about causal compressibility has not yet received much critical attention, although his work is cited favorably by other species selectionists, such as Jablonski 2008. And yet, his work is highly significant because it represents the most sophisticated attempt, to date, to spell out a notion of emergence that will do the work that species selectionists need.

Notice that causal compressibility is a feature of historical processes. Grantham's approach has the following noteworthy consequence. It is possible to imagine two populations, P1 and P2, which have exactly the same sex ratio at a given time. But we can imagine that P1 and P2 have very different histories. In P1's case, the individual-level processes leading up to P1 having that sex ratio are causally compressible. In P2's case they are not. Imagine that in P2's case, the sex ratio is not the result of any selective forces that are easy to describe, but rather is the result of an extraordinarily complex series of largely accidental events and processes. So sex ratio is an emergent property (in Grantham's sense) of P2 but not of P1, even though the sex ratio of both populations is the same. This result is a consequence of Grantham's treating emergence as a historical concept. Some might find this counterintuitive. But that need not be a serious strike against the view. In philosophy as well as in science, the correct views may often be counterintuitive.

One potential objection against Grantham's proposal is that his sense of "emergence" still is not strong enough to do the work that species selectionists need. Suppose, for the sake of argument, that Grantham is right about geographical range size: it really is emergent in his sense. The historical processes leading to a species – say, Jablonski's Cretaceous mollusks – having a certain range size are causally incompressible. A skeptic about species selection may still complain that range size is fully constituted by the spatial properties of the individual organisms at a given time. Thinking synchronically for a moment, if you knew all the details about where the individual organisms are, you would know all there is to know about the range of the species. In that sense, geographical range size is still a kind of aggregate property. Grantham, of course, would acknowledge all of this; he is just seeking to articulate a sense in which a species-level trait can be emergent even though these things are true. The skeptic, however, might argue that Grantham's notion of causal incompressibility doesn't help much here. As long as the species-level trait is fully determined by the properties of the individual organisms, macro-level processes will simply be by-products of processes going on at the individual level. Range will always be a by-product of the locations and movements of individual organisms. That fact won't change depending on whether we are able to offer compressed descriptions of the lower-level processes. In reply, Grantham would surely argue that his whole point is that species selection for geographical range size can be a distinct macro-level mechanism *even though* geographical range size is a by-product of the locations, ambits, and movements of individual organisms. That is because range size is causally incompressible.

Or is it? One challenge for Grantham's view is figuring out how to tell, in any given case, whether the species-level property really is causally incompressible. Grantham's response to this challenge is perhaps not entirely satisfying. He argues that:

> Because range is a product of many factors that interact in non-linear and context-sensitive ways, the dynamics of range growth and contraction will generally be causally incompressible. (2007, p. 83)

The problem, in a nutshell, is that in most cases, range size is determined by the interaction of so many complicated factors that it won't be possible to tell a simple (compressed) story about how different microevolutionary forces lead to expansions and contractions of a species' range. For example, Grantham asks us to imagine a species that is expanding its range at a given rate. At the same time sea levels are rising. Rising sea levels cut off a land bridge, thus blocking the species' expansion in one direction. This barrier is an accident of historical timing. If the species had expanded just a bit faster, or if sea levels had risen just a bit slower, the species would have made it across the land bridge. The point is that species range is determined by a whole host of such factors. For that reason, there will not usually be any shortcut to explaining the current range size; we need to know all the details. However, Grantham also admits that "range may not always be weakly emergent at the species level" (2007, p. 83). He allows that there might be cases in which changes in range size are driven in by selective forces operating at the organismal level, so that range size could in fact be causally compressible. Range size, according to Grantham, is usually causally incompressible, but need not be so. One potential worry here is that the species-selectionist will now need to show, on a case-by-case basis, that geographical range (or any other species-level property of interest) is causally incompressible.

What should we make, at this point, of the overall case for species selection? I hope the discussion of Grantham's work has made it clear just how difficult it is to show that species selection is really occurring in nature when we adopt a strong conception of what species selection requires. Jablonski's work on Cretaceous mollusks comes close. He showed that geographical range size is heritable, that it matters for species-level fitness. However, Grantham's work suggests that if we want to make good on the idea that macroevolutionary selection for larger range sizes really comes apart from microevolutionary processes, we would also need to show that range size is causally incompressible. He does offer some reason for thinking that it is. But at the same time, it isn't obvious that we should just assume that the range sizes of Jablonski's marine mollusks are causally incompressible. Grantham's admission that range size could be causally compressible suggests that we need to approach this on a case-by-case basis. That will make it rather difficult to show that species selection is a very important evolutionary phenomenon that occurs in a wide range of cases.

The emergent fitness approach

According to the emergent character approach, species selection requires that species have emergent features that are (a) heritable, and (b) make a

difference to probabilities of extinction and/or speciation. It's entirely natural here to zero in on the question of emergence, because it's natural to
think that what really matters here is the relationship between the species-
level properties and the properties of the individual organisms. However,
some theorists have argued that it could be a mistake to get hung up on the
issue of emergence. Okasha (2006) endorses an "acid test" for species selection
that was originally proposed by Elizabeth Vrba (1989): species selection "must
in principle be able to oppose selection at lower hierarchical levels" (2006,
p. 207; Vrba 1989, p. 115; see also Grantham 1995). Okasha parts ways with
Vrba, however, in suggesting that this acid test could well be met even in
some cases where the relevant species-level traits are merely aggregate. Indeed,
Okasha is generally skeptical about the idea that species-level traits need to
be emergent in order for genuine species selection to occur. His more relaxed
view of what is required for species selection makes it easier to argue that
certain evolutionary case studies qualify as genuine instances of it. His view
is closer to that of Lloyd and Gould (1993a), who also argued that genuine
species selection can occur even if the species level traits that account for
fitness differences are aggregate.

Earlier I said that the most interesting cases of species selection would be
those in which the following four conditions are met:

1. Species sorting is biased;
2. The bias is due to differences in species-level traits;
3. Those species-level traits are heritable, in the sense that "parent" species
   pass them on to their descendants; and
4. The species-level traits that make the difference are *emergent*.

As we have seen, the fourth condition is the source of a great deal of trouble. It
is a major challenge to try to develop an analysis of "emergence" that is appropriate to the task. Then there is the further empirical challenge of showing
that condition 4 is actually met in any given case. In light of these difficulties,
it is worth asking whether condition 4 is really necessary. Richard Lewontin
(1970) famously argued that evolution by natural selection occurs wherever
there is heritable variation in fitness. That idea is roughly captured by conditions 1 through 3. Perhaps that is all we need. If we drop the requirement of
emergence, then *bona fide* episodes of species selection will be rather easier to
come by.

Okasha concedes that at the end of the day, any fitness differences between species – for example, differences in fitness caused by differences in geographical range size – will always be traceable to factors at the level of individual organisms. The crucial point, for him, is whether those differences in species-level fitness are traceable to the workings of selection at the level of individual organisms. Do the differences in species-level fitness merely reflect differences in the fitness of organisms? Okasha writes that:

> The emergent character requirement stems from conflating the question whether lower-level *selection* is responsible for higher-level character-fitness covariance, with the question whether *some lower-level processes or other* are responsible. The former, not the latter, is the salient question; for on plausible metaphysical assumptions the answer to the latter will always be yes. (2006, pp. 207–208)

If the species-level fitness differences merely reflect differences in the fitness of organisms, then we do not have genuine species selection. Okasha's point is that it is possible for fitness differences among species to be caused by aggregrate species-level traits without being reducible to fitness differences among organisms. Although Okasha himself doesn't put it this way, perhaps one way of understanding his view is to replace the emergent character requirement (condition 4 above) with something like the following:

> 4* Fitness differences among species (i.e., differences in probabilities of extinction, speciation) are not simply reflections of fitness differences at the level of individual organisms.

Condition 4* represents another way in which macroevolution might be decoupled from microevolution. Notice also that Jablonski's case of geographical range size could satisfy this condition rather easily. In that case, it's easy to imagine the individual organisms in a species with a smaller geographical range being extremely fit relative to their local environments. Condition 4* just says that species selection cannot occur where species-level fitness itself is a mere aggregate measure (say, an average) of the fitnesses of individual organisms. Species selection requires a failure of aggregativity for fitness itself. And that can happen even though fitness differences at the species level are due to species-level traits that are aggregative.

Okasha is relatively more optimistic about finding examples of genuine species selection in nature. Indeed, he claims that Vrba's case of biased sorting in African antelopes is a real case of species selection! Recall that in that example, wildebeests had a much higher speciation rate than impalas. Okasha observes that the difference in speciation rates – or the difference in emergent, species-level fitness – was not simply a reflection of fitness differences in the organisms. Indeed, the impalas, with their low probability of speciation, seem to be doing quite well, based on their relative abundance. Because this case satisfies condition 4*, Okasha argues that it is a real instance of species selection. Indeed, Okasha also endorses Stanley's idea that species selection has played a role in maintaining sexual reproduction (2006, p. 207).

Science and philosophy in the species selection debate

It should be apparent by now that the species selection debate is messy and convoluted. Much of the debate focuses on empirical case studies, but much of it also focuses on conceptual questions about what species selection requires. Theorists with broader *vs.* narrower conceptions of species selection can and do disagree about whether a particular case study is a *bona fide* instance of it. Obviously, the more cases of species selection one can point to, the more significant species selection becomes as an evolutionary mechanism. But at the same time, one can increase the frequency of *bona fide* cases of species selection just by adopting a weaker view of what species selection involves – and thus making the whole idea less theoretically significant in another way. Paleontologists are caught in this trade-off: they can purchase one kind of theoretical significance, but only at the expense of another. Much of the debate is really about how best to negotiate this trade-off. But the question of how best to negotiate a trade-off like this one is not an empirical question.

The stakes for paleontology are quite high. I argued in earlier chapters that although PE represents a major theoretical achievement, it alone does not require any departure from minimalist models of evolution. PE is perfectly compatible with the idea that macroevolution is fully reducible to microevolution. By contrast, the move toward species selection and a hierarchical expansion of evolutionary theory really does represent a major revision of mid twentieth-century ways of thinking about how evolution works. For scientists such as Jablonski, Stanley, Vrba, and of course Gould, defending species selection is a way of defending the relevance of paleontology to evolutionary theory. To a very large extent, paleontology's status as a science with a serious theoretical contribution to make depends on the fate of species selection.

The species selection debate bears a striking resemblance to traditional debates in philosophy, where philosophers might disagree about broader *vs.* narrower analyses of an important concept. In presenting that debate in this chapter, I have deliberately set it up as a case of conflicting analyses. Defenders of the emergent fitness *vs.* the emergent character approaches have a shared view of the necessary conditions for species selection but disagree about what additional conditions need to be met. In addition, defenders of the emergent character approach have internal issues about how to analyze the concept of emergence. This is what debates in philosophy often look like. Two camps of philosophers will agree about the necessary conditions for something – say knowledge – but they will disagree about what additional conditions need to be met. Furthermore, philosophers will often fixate on cases in which the opposing camps disagree about whether the concept applies – for instance, a case in which one camp says that Smith has knowledge, while the other camp denies that. Again, the species selection debate has this flavor: those who favor the broader *vs.* the narrower conceptions of species selection disagree about how to classify cases, such as Vrba's African antelope case. The parties to the debate tend to fixate on those problem cases, precisely because those are the ones that bring out the theoretical disagreements. The paleontologists have done a great deal of philosophizing.

The meta-level observation that the species selection debate looks like a philosophical one raises the following question: Is it generally true of science that if you work from the bottom up and doggedly pursue the theoretical questions, you will eventually find yourself in philosophical territory where the issues are largely conceptual and only loosely constrained by the empirical data? Or have we discovered something special about paleontology? Is there something about the difficulty of testing our ideas about evolutionary history that gives macroevolutionary theorizing more of a philosophical tint? These questions are too large for me to try to answer here in a responsible way, but I will say that I suspect that the former is true, and that what we have discovered here is a deep commonality between paleontology and other areas of science: if you start at the bottom and work upwards, things will eventually become more philosophical. This discovery, perhaps more than anything else, shows that paleontology isn't just about collecting and displaying fossils. It deserves a place among the most theoretical of natural sciences.

6   Real trends, relative progress

"Progress is a noxious, culturally embedded, untestable, nonoperational, intractable idea that must be replaced if we wish to understand the patterns of history"

           – Stephen Jay Gould (1988a, p. 319)

"It is reasonable to ask whether the fossil record shows evidence of successively 'better' organisms through geological time"

           – David Raup (1988, p. 293)

Evolutionary paleontologists often think of their work as involving a kind of two-step procedure: they begin by identifying and documenting patterns in the fossil record, then they try to draw inferences about the underlying evolutionary processes that give rise to those patterns. The terms "pattern" and "trend" are not exactly interchangeable, but they are close. A trend is one kind of pattern. Scientists standardly define a trend as any persisting directional change in some interesting variable over time. The debates concerning PE and species selection are really debates about patterns and trends. For example, PE involves a claim about how to see certain patterns in the fossil record; while species selection involves a claim about the processes that give rise to such patterns. At this point in the book, I want to throw the door open to some deeper philosophical questions about patterns and trends.

In this chapter, I explore two preliminary questions about trends. First, what does it mean to say that a trend is or isn't real? And second, how is the study of trends related to questions about evolutionary progress? In the next chapter, I consider just how much science can really tell us about the causes of historical trends.

Before going any further, though, it is worth emphasizing that some questions about historical trends go way beyond paleontology. All of the following are examples of putative historical trends:

Falling housing prices

Directional change in gene frequencies in a population

Human population growth

Global warming

Increasing hurricane intensity

Grade inflation

The westward drift of the mean geographical center of population of the US

Increasing number of children raised in single-parent households

Decreasing unemployment

Eye reduction and loss in trilobites during the Devonian period

Increasing biological diversity

Documenting and explaining historical trends is a central project in economics, sociology, and some of the other social sciences, not to mention the
study of human history. In some areas of research, such as climate science
and economics, theorists do their best to try to project trends into the future.
Anyone interested in questions about trends would actually do well to focus
on paleontology. In one sense, paleontologists' work on large-scale evolutionary trends has lower stakes than work on historical trends in economics or
climate science. No policy decisions hang on what the paleontologists tell us
about trends in the evolution of complexity or of body size. It can sometimes
be easier to see things clearly when the stakes are lower and when fewer people
have a vested interest in the outcome of an investigation. Paleontologists have
also made a great deal of progress in developing methods and concepts that
enable them to study historical trends in a rigorous and quantitative fashion.

Is Cope's rule real?

Cope's rule, named in honor of the nineteenth-century American paleontologist E.D. Cope, says that body size usually increases over the course of
evolutionary time. Cope himself never actually formulated the rule that has
come to bear his name (Polly 1998). Cope (1896) did introduce the closely
related "law of the unspecialized," which says that the dominant organisms
of a later epoch are usually descended from the ecological generalists of
an earlier period. In addition to being a rather speculative neo-Lamarckian
macroevolutionary theorist, Cope was also a dinosaur scientist, and he is perhaps best known for rivalry with Yale paleontologist O.C. Marsh. Cope and

Marsh engaged in a fierce competition to see who could retrieve the most dazzling dinosaur skeletons from what was then still a rather wild west (Jaffe 2001). Cope had mammals and dinosaurs in mind when he formulated the law of the unspecialized. He knew that the mammals of the Mesozoic were all rather small ecological generalists. Those generalists, rather than the highly specialized dinosaurs, went on to become the dominant group of the Cenozoic era. Cope's idea was that this might reflect a more general truth about evolution.

Cope's law of the unspecialized was one of a proliferation of so-called "laws" of macroevolution that scientists proposed in the late nineteenth and early twentieth centuries. Another example was Dollo's law – first proposed by the Belgian paleontologist Luis Dollo – according to which evolutionary processes are irreversible. Dollo argued that once species B evolves from species A, B will never evolve back into A. Evolution never retraces its steps. Dollo's law is related to ideas about irreversibility in thermodynamics and other areas of science. In addition to the law of the unspecialized and Dollo's law, some scientists began to talk about "Cope's law" of size increase.

The term "Cope's law" fell out of fashion some time ago, and there are a couple of reasons for that. For one thing, scientists found that the fossil record contains some clear exceptions to the law of size increase. For instance, horses were long touted as a group characterized by steady, gradual size increase. Horses evolved from a small dog-sized animal called *Hyracotherium* that lived some 60 million years ago. More recently, though, there was one horse lineage that underwent a dramatic evolutionary size reduction (MacFadden 1986). There is also a well-known evolutionary phenomenon called island dwarfism. Sometimes when a population gets stranded on an island, it ends up going through a radical evolutionary size decrease. Fossil remains of dwarf elephants have been found, for instance, on the Mediterranean islands of Cyprus and Malta, and remains of dwarf woolly mammoths have been found on islands in Alaska. The so-called hobbit fossils – small and surprisingly recent hominins found in a cave on the island of Flores, in Indonesia – may also be an example of island dwarfism. Now a law of nature is supposed to be exceptionless, and to hold true in all times and places. These and other well-documented exceptions provide good reason for demoting Cope's law to the status of a mere rule or statistical generalization. Most paleontologists today take Cope's rule to say that size *usually* increases over evolutionary time. (Note the parallel with the debate concerning PE. Proponents of PE claim only that evolution is usually punctuational.) Indeed, the consensus today is that there is really no point in searching for genuine laws of large-scale evolutionary change, as long as laws are thought to be true universal generalizations.

Is Cope's rule a real trend? We know that the earliest forms of life on Earth were very small. From the first appearance of life some 3.8 billion years ago until the first multicellular life, which evolved a little more than a billion years ago, the biggest organism on the planet was a single cell. Although even today much of the world's biological diversity is microbial, the largest organisms are orders of magnitude bigger than the biggest organisms of the Earth's early history. At the very least, we have here an example of what Shanahan (2004) helpfully calls an "apex directional trend," or a trend in the maximum body size of living things. Mammals, too, exhibit an apex directional trend toward larger body size. 65 million years ago, the largest mammals were quite small, probably the size of small dogs. But today, the largest mammals are vastly bigger. However, when scientists talk about Cope's rule, they typically do not have in mind a trend in the maximum; rather, they have in mind a trend in the mean.

A trend in the maximum might not tell us anything very interesting about evolution. It is plausible to suppose that there is a minimum size for living things made out of the basic chemical building blocks available on our planet. Imagine taking a single-celled prokaryotic organism and trying to shrink it. Beyond a certain point, it would surely get so small that it could no longer carry out the basic chemical processes associated with life. Facts about biochemistry impose a constraint or lower limit on the size of living things, and life on Earth started very near that lower limit. That means that if life diversifies at all, there will be an apex directional trend toward larger size. The increase in the maximum body size could simply be a by-product of increasing diversity, a point that Gould has emphasized (1988b; 1996). As long as our evolutionary theory can explain the diversification, there is no need for any further explanation of the size increase. A trend in the mean body size of an evolving group has the potential to be more interesting, although even that is disputed.

In 1997, Gould published an article in the prestigious journal, *Nature*, entitled "Cope's rule as a psychological artefact." He argued that Cope's rule is not a real trend, but rather that it is a kind of mirage that results from the ways in which scientists look at the fossil record. That paper contains two distinct lines of attack against Cope's rule.

First, Gould pointed out that in a much earlier paper, Steven Stanley (1973) had argued that Cope's rule could result from random processes. Stanley was responding, in particular, to even earlier writers, such as Newell (1949), who had suggested that Cope's rule is best explained in terms of natural selection. Perhaps natural selection, for whatever reason, favors organisms with larger body sizes. All it takes is a little imagination to see why larger organisms might thereby gain an edge in the Darwinian struggle for existence: they might be more difficult for predators to kill; they might have an easier time withstanding food shortages; they might be more attractive to potential mates; they might be better able to defend their territories; they might live longer; and so on (see e.g., Hone and Benton 2005). Stanley identified two very serious problems for this "bigger is fitter" view. First, he argued that it makes sense to think about natural selection as pushing a population from a starting mean body size toward an optimum body size for whatever niche that population happens to occupy. On this view, natural selection doesn't favor larger body size, *per se*; rather, it favors larger body size in those cases where the starting mean body size of the population is well below the optimum. This makes sense: if a population started out with a mean body size far larger than the optimum size for the niche it occupies, then natural selection would lead to size decrease. Now suppose that Cope's rule is the norm. That could only mean that populations typically start out with mean body sizes well below the optima for their respective ecological niches. Why would that be the case? Fascinatingly, at this juncture, Stanley invoked an old idea from E.D. Cope himself, which was closely related to Cope's law of the unspecialized. In effect, Stanley used an idea of Cope's own to attack the "rule" which had (probably mistakenly) come to be attributed to Cope. Cope suggested that having a large mean body size might make a species more prone to extinction – an early version of the concept of biased species sorting. Again, Cope was probably thinking of the dinosaurs and the mammals here. He knew that at the end of the Cretaceous, all the large-bodied dinosaurs became extinct, while the relatively smaller mammals persisted. This idea has recently been challenged by the discovery of some "giant" gastropod fossils from the period immediately following the massive Permian-Triassic extinction, when one would think that the larger gastropods would have died out (Brayard et al. 2010). Still, the idea that larger body size increases extinction risk has some intuitive plausibility.

Drawing inspiration from Cope, Stanley argued that what really needs explaining is why lineages typically start out so small, and that the best explanation for that so-called "Lilliput effect" is biased species sorting. Stanley called his own explanation of Cope's rule a "probabilistic" one. His point was that if lineages start out small, at or near the minimum size constraint for organisms of that type, then chances are their body size will be smaller than the optimum. As the lineages in the clade diversify, size optima might shift around, and new ecological niches will open up. The result will be a gradual drift away from the lower size limit. Stanley's main take-home message was that if you want to explain Cope's rule, you need to explain why things start out small.

Stanley's argument on this score is extremely important, but it's not clear that this argument gives us any reason to doubt the reality of Cope's rule. Indeed, in his 1973 paper, Stanley grants that Cope's rule is a genuine phenomenon; he is more interested in explaining the causes of Cope's rule, and in challenging the traditional, simplistic explanation in terms of natural selection. Notice, though, that Stanley does also leave a role for selection to play. Indeed, he wrote that "certainly when evolutionary size increase occurs, it is in response to selection pressure resulting from one or more advantages" (1973, p. 1). In his picture, selection is still needed to drive lineages from their starting size toward the optimum. What he objected to most strenuously is the idea that bigger is necessarily fitter.

Stanley's 1973 paper happened to coincide with the work of the MBL group, discussed in Chapter 4. Recall that the MBL scientists were using an early computer simulation – the so-called MBL model – to represent phylogenetic branching processes over time. With each time interval, or with each "turn" in the simulation, a lineage has a specified probability of persisting with no change, branching, or going extinct. The MBL group also used their simulation to model morphological change, such as change in body size (Raup and Gould 1974). All one needs to do is to specify a starting mean body size for the lineage, and then stipulate that body size has a certain probability of increasing and of decreasing with each time interval. If you want to simulate the action of natural selection on the lineage, then all you need to do is to introduce a directional bias. You could tell the program that size increases are much more probable than size decreases; that would represent selection's favoring larger body size. The MBL scientists showed that it is possible to get the computer program to generate an upward trend in the mean even if you stipulate that size increases and decreases are equally probable, or random. That is, you can generate a trend even if you remove natural selection from the picture

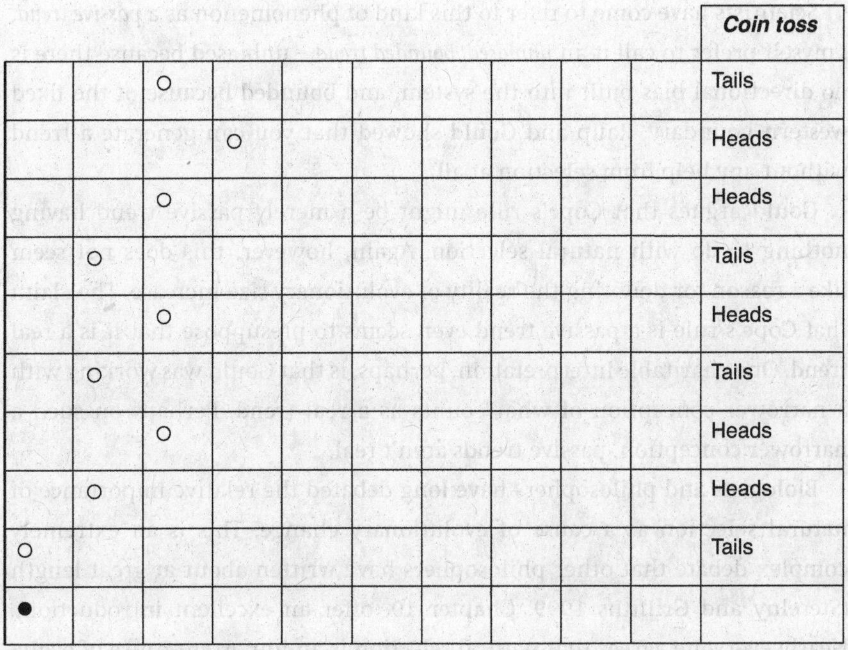

										Coin toss
	○									Tails
		○								Heads
	○									Heads
○										Tails
	○									Heads
○										Tails
	○									Heads
○										Heads
●										Tails
●										

Figure 6.1 A model of a random walk away from a fixed boundary. The left-hand side of the grid represents a fixed boundary. Heads means the token moves forward and one square to the right. Tails means forward and one square to the left.

entirely. The trick is just to make sure that the evolving clade starts out at or near the minimum size constraint. It will then do what is known as a "random walk" away from that fixed lower boundary.

The concept of a random walk is easily illustrated by means of a simple coin tossing game. Consider the set-up shown in Figure 6.1, with a token at the lower left-hand side of the board. With each turn of the game, the token advances one step upward. Then you flip a fair coin. If the coin comes up heads, the token moves one step to the east; if tails, one step to the west. If the coin comes up tails on the first turn, it merely advances one square to the north, because the western boundary constrains it. If you ran this simple game over and over again, you would find that time after time the token does a random walk away from the fixed western boundary. It might help to think of the token as representing a single lineage. The simulation that Raup and Gould used was like this one, but with the possibilities of branching and extinction built in.

Scientists have come to refer to this kind of phenomenon as a *passive trend*. I myself prefer to call it an *unbiased, bounded trend* – unbiased because there is no directional bias built into the system, and bounded because of the fixed western boundary. Raup and Gould showed that you can generate a trend without any help from selection at all.

Gould argues that Cope's rule might be a merely passive trend having nothing to do with natural selection. Again, however, this does not seem like a reason for doubting the reality of evolutionary size increase. The claim that Cope's rule is a passive trend even seems to presuppose that it is a real trend. One charitable interpretation, perhaps, is that Gould was working with a narrower conception of what counts as a real trend. Perhaps on such a narrower conception, passive trends aren't real.

Biologists and philosophers have long debated the relative importance of natural selection as a cause of evolutionary change. This is an extremely complex debate that other philosophers have written about at great length (Sterelny and Griffiths 1999, Chapter 10, offer an excellent introduction). Nearly everyone agrees that natural selection is an important cause of evolutionary change. At the same time, everyone agrees that there are other factors, aside from selection, that can make a difference in evolution: mutation, migration, random genetic drift, etc. Everyone also agrees that selection must operate within certain biological constraints. The main point of contention in this debate is how important natural selection is *vis-à-vis* these other considerations. On the one hand, *adaptationists* of various stripes tend to play up the importance of natural selection. Without denying that the other factors often matter, adaptationists tend to see natural selection as the main driving force behind most evolutionary change. But on the other hand, many other scientists and philosophers of science have criticized adaptationism and sought to downplay the importance of selection. Chief among these critics is Stephen Jay Gould, whose famous co-authored essay, "The Spandrels of San Marco and the Panglossian Paradigm," is a classic anti-adaptationist manifesto (Gould and Lewontin 1979). Many of the theoretical arguments that Gould made in paleontological contexts similarly involve a downplaying of the importance of selection, although one sometimes has to read between the lines to see this (Sterelny 2001 provides an excellent overview). Gould's claim that Cope's rule might be a passive trend is a case in point. Although not all paleontologists share Gould's opposition to adaptationism, Gould himself certainly thought that one way in which paleontology can contribute to evolutionary theory is by showing biologists that natural selection is not as important as many have thought.

Another major issue in the philosophy of biology that comes into play here is the role of chance in evolution. Most biologists take the view that chance does have some role to play in microevolution. For example, the standard view is that mutation and random genetic drift represent (though in somewhat different ways) chance factors in microevolutionary change. Most would also say that it is an empirical question just how much evolutionary change is due to chance *vs.* natural selection. In suggesting that Cope's rule might be a passive trend, Gould is, in effect, suggesting that one of the most significant, large-scale patterns in the history of life on Earth is due to chance. It's not entirely due to chance, because a passive trend always requires some kind of fixed boundary, in this case, a fixed lower limit on body size, and that fixed boundary is not simply there by chance. Still the general point is that although Cope's rule may seem like a parochial issue within paleontology, Gould's attack on it raises fundamental questions about how evolution works.

The second prong of Gould's attack on Cope's rule was empirical. He trumpeted a study by David Jablonski (1997) that was published in the same issue of *Nature*. Jablonski's work represented one of the most serious attempts up to that time to put Cope's rule to the empirical test. Recall from Chapter 5 that Jablonski had studied marine bivalves and gastropods from the Cretaceous period in order to see whether geographic range size correlates with extinction risk. He used roughly the same group of organisms to test Cope's rule. In his 1997 paper, he looked at the body sizes of 1,086 species of Cretaceous mollusks, occurring in 191 distinct evolving lineages, in order to see whether body size increase was the norm. The specimens he studied all came from a 16-million-year time interval. Jablonski found that less than one third of these mollusk lineages (27–30%) exhibited within-lineage size increase, and that almost as many (26–27%) showed size decrease. Gould saw Jablonski's work as undermining Cope's rule:

> The obvious test requires that we abandon our habit of selective search for the
> expected and, instead, study all lineages in large clades with excellent data
> over substantial geological intervals. Jablonski has followed this admirable
> procedure in the most comprehensive set of data ever assembled to test Cope's
> rule – and the rule fails in this case. (Gould 1997a, p. 199)

Gould combined this empirical argument with a point about human psychology. He suggested that we all think, at some level, that "bigger is better," and so we naturally fixate on those lineages and clades that do exhibit size increase. We mistakenly take those to be representative cases, and then proceed to draw erroneous conclusions about how evolution works in general. This is what Gould means when he says that Cope's rule is a "psychological artefact." The fossil record contains a staggering amount of variety. If scientists hope and expect to see evolutionary size increase, then it will be easy for them to attend selectively, perhaps without quite realizing what they are doing, to series of fossils that do exhibit size increase. Paleontologists, like everyone else, are subject to confirmation bias. To make this point vivid, think about Jablonski's work: he did in fact find that many lineages of Cretaceous mollusks exhibit size increase. What if he had studied just a subset of those lineages? He would have found that Cope's rule holds true.

In at least one respect, Gould's argument here is a bit of a stretch. It's a little hard to believe that when it comes to body size, scientists have the general prejudice that bigger is better. That view would commit one to saying that the sauropod dinosaurs were the most impressive land animals that ever lived. The largest organisms alive today are probably plants and fungi. For example, a whole stand of aspen trees connected by underground roots may count as a single organism. When Gould attributes to other scientists an unconscious bias or preference, he is in effect making an empirical psychological claim, and the onus is on him to provide some evidence for it. Even worse, the point about bias cuts both ways. Some scientists might have strong motivations for questioning Cope's rule. Those motivations might lead them to attend selectively to clades and to timeframes in which Cope's rule seems not to hold up. If it's possible to challenge research supporting Cope's rule by alleging bias, then one can challenge research undermining Cope's rule in exactly the same way.

Although Gould's argument fails to undermine Cope's rule, it does bring out one interesting fact about the reality of historical trends – namely, their dependence on scale. Whether a trend shows up as real depends on the prior decisions that one makes about which clade(s) to study, and also which time intervals to focus on. Those prior decisions make all the difference to what one sees. One can reasonably read Gould here as raising a general concern about how paleontologists make those decisions about what to study. Those decisions may be informed by empirical considerations without being

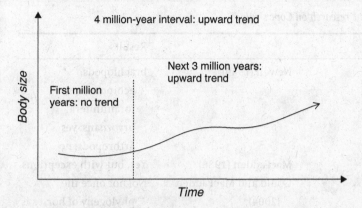

Figure 6.2 The relativity of trends to time intervals.

dictated by them, and there is surely room for various kinds of biases to creep in.

In order to appreciate the way in which trends are subject to scaling effects, consider a simple illustration. Imagine a single lineage whose mean body size evolves as depicted in Figure 6.2. If you were to look at the whole time interval – say, 4 million years – you would see an upward trend in body size. But if you only looked at the first million years of this lineage's history, you would see stasis. Whether size increase is a real trend depends on the time interval one looks at.

The reality of the evolutionary trend can also depend on the clade one decides to look at. Imagine a case where a clade of marine mollusks, say, is undergoing steady evolutionary size increase over a period of several million years. But that clade also belongs to a larger clade that is experiencing no increase in the mean body size at all. The difference could be due to a number of factors. For example, it could be that other lineages are getting smaller, or that there is some species sorting taking place. The point is just that Cope's rule will show up as a real trend in the smaller sub-clade, but not in the larger clade, even over the same time interval. What this means is that there is no simple, easy answer to the question whether Cope's rule is a real trend. The only answer that makes any sense at all is one that sounds relativist: it depends.

This relativist view may seem pessimistic, but it actually helps to make sense of the lively paleontological research program on Cope's rule. Since the mid 1990s, more and more researchers have been doing just what Jablonski did in the paper that Gould celebrated: they are tackling the problem of

Table 6.1 *Recent research on Cope's rule.*

Target Group		Results
Invertebrates	Newell (1949)	brachiopods, echinoderms, foraminifera, bryozoans: yes arthropods: no
Horses	MacFadden (1986)	Yes, but with exceptions
	Gould and MacFadden (2004)	No, not once the phylogeny of horses is properly clarified
Cenozoic foraminifera (plankton)	Arnold, Kelly, and Parker (1995)	Yes
	Schmidt, Thierstein, and Bollman (2004)	Depends on the time interval
Cretaceous mollusks	Jablonski (1997)	No
Mammals	Alroy (1998)	Yes
Early Jurassic ammonites	Dommergues, Montuire, and Neige (2002)	No
Extant North American freshwater fish	Knouft and Page (2003)	No
Early amniotes of the Devonian and Permian	Laurin (2004)	Reptiliomorphs: Yes Stegocephalians: No
Recent North American canids	Van Valkenbergh, Wang, and Damuth (2004)	Yes
Dinosaurs	Hone et al. (2005)	Yes
Deep sea ostracodes	Hunt and Roy (2006)	Yes
Pterosaurs	Hone and Benton (2007)	Yes, but with varying results at different taxonomic levels
Mesozoic birds	Hone et al. (2008); Butler and Goswami (2008)	Contested

documenting Cope's rule in piecemeal fashion, by looking at this or that clade over this or that time interval. Some of the results of this research are summarized in Table 6.1. The idea that what you see depends on the clade and the time interval you focus on is just a background assumption of this entire line of research. No one thinks that a negative finding will somehow "refute" Cope's rule, or that a positive finding will enshrine it once and for all. The general consensus seems to be that we need lots of studies, all conducted in the spirit of Jablonski's work, in order to get an increasingly detailed documentation of trends in the fossil record.

Why have paleontologists expended so much energy on the study of body size evolution? One reason is that body size is relatively easy to define. Increasing complexity is another example of a large-scale trend in evolution that has interested many scientists, but it is difficult to provide a working definition of complexity that would apply to all different kinds of living things (McShea 1991). One traditional definition of biological complexity focuses on number of cell types – skin cell, muscle cell, nerve cell, blood cell, and so on. But this approach doesn't help us understand differences in complexity among single-celled organisms. Nor does the fossil record always contain clear information about the number of cell types an organism has. Body size is much easier to define and much easier to estimate from the fossil record. Often it is even possible to estimate body size from incomplete fossil remains. From an evidential point of view, body size seems to represent the best-case scenario for the study of evolutionary trends. And although it may seem like a trivial observation, body size is also a feature that every single living thing happens to have; this means that studying body size is a good way to obtain the big picture of evolutionary history. One could, after all, study the evolution of brain size (and some scientists do), but since many living things lack brains, studying trends in brain size will not provide the big picture. Paradoxically, another reason for focusing on body-size evolution, as opposed to a feature such as complexity, may be that the stakes are lower. Since no one has ever seriously proposed that body size increase constitutes evolutionary progress, no one has much of a vested interest in the outcome of the investigations. Then again, some scientists treat body size as a kind of proxy variable for complexity. Increases in complexity seem to require increases in body size. This connection to complexity makes body size more interesting than it might seem at first.

Trends and patterns as abstract objects

The research program depicted in Table 6.1 is aimed at determining, albeit in piecemeal fashion, whether Cope's rule is a real trend. The jury on the larger question is still out, but with each new study that is published, scientists' overall picture of evolutionary history becomes more and more complete. One might still wonder, though, what exactly it means to say that a trend is real at all. With one noteworthy exception – an essay called "Real Patterns" by philosopher of mind Daniel Dennett (1991) – philosophers have had very little to say about the metaphysics of trends. Dennett himself doesn't even mention paleontology in the "Real Patterns" essay, but his work contains insights that can be applied in the philosophy of paleontology.

One of the most heated debates about the reality of a pattern in the fossil record was the so-called Nemesis affair (Raup 1985). In the early 1980s, Jack Sepkoski and David Raup looked at vast numbers of fossils from the last 250 million years, and they found an intriguing pattern. It looked like mass extinction events took place on a regular schedule, approximately every 26 million years (Raup and Sepkoski 1984). This mass extinction periodicity was a stunning result because there is absolutely nothing at all in the modern synthetic theory of evolution that provides the slightest reason to think that mass extinctions should be cyclical. The result looked like a potential triumph for the new statistical paleobiology. If it held up, then it would decisively show that the fossil record has things to teach us about how evolution works that we could not discover in any other way. Not only that, but the 26-million-year extinction periodicity also looks like a macroevolutionary pattern that simply cannot be reduced to microevolutionary processes. This evidence therefore posed a serious challenge to minimalist models of evolution. In short: Raup and Sepkoski's work on extinction periodicity had pretty much everything that the new paleontologists were looking for in their attempt to establish paleontology as a serious quantitative discipline with something to say about evolutionary theory.

Almost immediately, other scientists began to wonder whether the 26-million-year cycle was indeed a real pattern. Many worried that the pattern was a mere illusion generated by the techniques that Raup and Sepkoski were using to study the fossil record. Raup and Sepkoski focused on the appearance, duration, and disappearance of families in the fossil record, rather than genera and species. They measured mass extinction events, for instance, by counting the number of families that disappeared at the same time. There are some good arguments in favor of that approach. There are probably many species that are never preserved in the fossil record, so the species that are preserved in the record represent a limited sample that is going to be biased in various ways, and it takes a lot of work to try to identify and correct for those biases.

As you move up the taxonomic hierarchy, from species to genera, to families, and beyond, the probability that a taxon will not be represented in the fossil record at all goes down dramatically. Focusing at the family level represents a reasonable compromise: you can still get a relatively high-resolution view of extinction patterns, but at the same time, you do not have to worry quite as much about sampling biases in the fossil record. Some of Raup and Sepkoski's critics quickly raised doubts about their approach. Patterson and Smith (1987; 1989) wondered if the 26-million-year extinction periodicity might have more to do with how scientists classify families than with anything else. If scientists used different methods of classification, would the extinction periodicity disappear?

One question on everyone's minds was why mass extinction events should occur on a 26-million-year schedule. Raup and Sepkoski just happened to publish their work at a time when the scientific community was intensely focused on the then-recent discovery of geological evidence pointing to an extra-terrestrial impact about 65 million years ago, right at the time of the Cretaceous-Tertiary mass extinction (Alvarez et al. 1980). The father and son team of Luis and Walter Alvarez had discovered deposits of iridium right at the K-T boundary, and iridium is extremely rare in the earth's crust. It exists in larger quantities deep inside the earth, and has also been known to occur in asteroids. In the early 1980s, the scientific community was working hard to try to discriminate between different hypotheses about the origins of the iridium in the K-T boundary clay: volcanism *vs.* asteroid impact. Many had begun thinking seriously about asteroid impacts as causes of mass extinction. Raup and Sepkoski hinted that there might be an extra-terrestrial cause of the 26-million-year extinction periodicity, and some astronomers ran with the idea. They posited the existence of a heretofore unobserved star, which they called Nemesis, whose pathway through the galaxy brings it past our solar system every 26 million years. Nemesis's gravitational field would then dislodge asteroids from their orbit around the sun and send some of them hurtling in our direction. While the debate about the reality of the 26-million-year pattern continued, subsequent observations utterly failed to detect Nemesis. Today, no one in the scientific community believes that Nemesis really exists.

The debate concerning extinction periodicity is a fascinating case in which scientists investigated the reality of two very different sorts of things. On the one hand, the paleontologists focused on the reality of the 26-million-year pattern; on the other hand, the astronomers focused on the reality of our sun's unwelcome twin star, Nemesis. For present purposes, I mainly want to contrast these two scientific debates about what's real.

To begin with, note that Nemesis, like any star, is (or would have been) a *concrete object*. I have no particularly good analysis of concrete objects, but it might be enough to define concrete objects ostensively by pointing to familiar examples. In addition to planets and stars, ordinary physical objects such as tables, chairs, rocks, and coffee mugs are concrete. But so are trees, animals, individual cells, and even atoms and molecules. So are triangular pieces of paper. Concrete objects are particular things that exist in space and time, and we can causally interact with them. Perhaps the best examples of *abstract objects* are geometrical shapes and numbers. Dennett (1991) suggests that centers of gravity are abstract objects, as is the mean geographical center of population of the US. Consider the number two, as distinct from the numeral "2," or from any pair of concrete physical objects. If the number two is a real thing – and that is much disputed – then it would seem to exist in some supernatural Platonic realm, outside of space and time. Some philosophers with Platonist leanings think that numbers really exist, while others have denied this and gone to great lengths to show how mathematics could make sense even if there are no numbers. Another interesting tradition in philosophy denies that there are abstract objects but allows that we have abstract ideas, or concepts. Still other philosophers, such as the eighteenth-century empiricists George Berkeley and David Hume, even deny that we have any abstract concepts. In short, abstract objects have been the source of major headaches in philosophy.

Trends and patterns seem like abstract objects (Dennett 1991). Cope's rule is usually understood to be a trend in the mean body size, or, if you will, a directional change in the value of the mean. But the average body size of all the sauropod dinosaurs at a give time is not a concrete particular thing; it's just a number. The 26-million-year extinction periodicity that Raup and Sepkoski reported is likewise merely a pattern, as opposed to a concrete particular thing, event, or process. It is relatively easy to understand what is at stake in an investigation into the existence or non-existence of some concrete object, such as a star. Disputes about the existence or non-existence of trends and patterns are far messier and more difficult to understand.

There is a deep metaphysical question about whether any abstract objects exist at all. That simply cannot be what the paleontologists are interested in. When Gould, for instance, questions the reality of Cope's rule, he is not saying that Cope's rule does not exist because there are no abstract objects. Dennett (1991) argues that the metaphysical issue is not really the interesting one. Consider the case of centers of gravity in physics. Some would say that centers of gravity are useful fictions. They don't really exist – because no abstract objects exist – but it is still helpful for physicists to think and talk about them. Others would say that centers of gravity are real, that they are just spatial points that we define a certain way. Dennett observes that philosophers with very different views about the deep metaphysical issues seem to agree about the usefulness of centers of gravity. With this in mind, he introduces an alternative, pragmatic way of thinking about the reality of abstract objects:

> Centers of gravity are real because they are (somehow) *good* abstract objects. They deserve to be taken seriously, learned about, used. If we go so far as to distinguish them as *real* (contrasting them, perhaps, with those abstract objects which are bogus), that is because we think they serve in perspicuous representations of real forces, "natural" properties, and the like. (Dennett 1991, pp. 28–29)

On Dennett's view, when you say that some abstract object X is real, you are in effect saying that X is useful for purposes of representing something else, or perhaps for making predictions. He thinks that making predictions generally involves recognizing patterns and projecting them into the future.

Dennett contrasts centers of gravity with other less useful abstract objects, such as the median geographical center of population of the US. The median center of population lies at the intersection of a line of latitude drawn such that half the population of the US is to the north, and half to the south, with a line of longitude drawn such that half the population is to the east, and half to the west. You might think that the median center of population is useful to demographers interested in tracking the westward migration of people across North America. But what about the schmedian center of population, which lies at the intersection of a line of latitude drawn such that 51% of the population is to the north, and 49% to the south, with a line of longitude drawn such that 51% live to the east and 49% to the west? Or what about the geographical center of your own lost socks (Dennett's example)? There is simply no end to the abstract objects we might define, and the vast majority of them are silly and uninteresting. Similarly, the only limit to the trends that one could track in evolutionary history is the imagination. Many of those will seem like matters of idle curiosity. Was there a directional trend in neck length in Jurassic sauropod dinosaurs? How about a trend in the ratio of neck length to tail length?

Dennett's pragmatist proposal for how to think about what is involved in saying that a trend or pattern is real confronts a couple of problems. To begin with, Dennett closely associates usefulness with prediction, but paleontologists typically do not use evolutionary trends to make predictions about the future course of evolution. We would not, however, want to say that all of the trends and patterns that paleontologists study are for that reason bogus. Still, Dennett's proposal might work here if we think of usefulness in a broader way. The deeper problem with Dennett's approach, it seems to me, is that the useless patterns and trends are still real. We can, for example, be wrong about them. Suppose that ratio of neck length to tail length of sauropods actually decreased during the Jurassic. If that were so, and if you hypothesized a persistent increase in that measure, then you would be wrong, plain and simple. Increase in the ratio of neck length to tail length would not (on this assumption) be a real trend, but a decrease in that measure would be. Even among the bogus, useless trends, we'll want to draw a distinction between the real trends and those that aren't. That alone reveals the mistake in trying to define "unreal" as "useless."

Later on in his 1991 paper, Dennett offers a second, potentially more useful way of thinking about what is involved in claims about the reality of patterns. He writes that:

> A pattern exists in some data – is real – if there is a description of the data that is more efficient than the bit map, whether or not anyone can concoct it.
> (1991, p. 34)

Notice that there is no reference at all to usefulness here. We might do better to call this the *informational approach*. Figure 6.3 shows a series of bar codes that a computer program generated by printing 10 rows of dots with 90 dots in each row: 10 black, 10 white, 10 black, and so on. The same pattern is repeated six times. Then the computer went back and added a certain amount of noise to each pattern.

A: 25%

B: 10%

C: 25%

D: 1%

E: 33%

F: 50%

Figure 6.3 Dennett's bar codes. Reproduced from Dennett (1991) with permission from *The Journal of Philosophy*.

The noise ratio in F is so high that the original pattern is impossible to make out. F could just as easily have been generated by a program printing black and white dots at random.

Next, Dennett asks us to imagine a situation in which we have to send information about frames A through F to someone else. You might describe frame B by saying that it consists of 10 rows of 90 dots each, with alternating sets of 10 black and 10 white dots in each row, but with certain exceptions. You could then describe the 10% noise pixel by pixel. That is a compressed description of frame B. If, however, you had to describe frame F to someone else, you could not offer any such compressed description; you would have to give the entire bit map, dot by dot. According to this informational approach, a real pattern exists in some data if and only if it is possible to offer a compressed description of the data. It also does a good job making sense of some of the examples we have considered. The US Census Bureau tracks the mean geographical center of population (which is defined a bit differently than the median center of population). Since the 1790s, the mean center of population has drifted westward from Maryland to Missouri. Talking about the westward drift in the mean is a way of offering a compressed description of a vast number of births, deaths, and relocations that have occurred over the last two centuries. Likewise, talking about a trend in the mean body size of a clade is a way of offering a compressed description of a vast number of events and processes: births, deaths, speciation and extinction events, and so on.

This informational approach helps to illuminate one other interesting difference between debates about the reality of some trend, such as Cope's rule, and debates about the reality of some concrete particular, such as the Nemesis star. When scientists posit the existence of some concrete particular thing or event that no one has yet observed, they typically do so with the aim of providing a causal explanation of something that we have observed. Thus, the Nemesis star was supposed to help explain other observations, especially the alleged periodicity of mass extinction events. However, when scientists posit the reality of a trend or pattern, they are doing something fundamentally different. Their project is a *descriptive* rather than an explanatory one. The trend or pattern is itself something that stands in need of explanation. Just how scientists go about explaining the occurrence of trends will be the main theme of the next chapter.

Before moving on, we should examine a couple of other interesting consequences of the informational approach. First, a single set of data can contain lots and lots of real patterns, in this informational sense. For example, the same set of census data will contain a trend in the median center of population, but also a trend in the schmedian, and plenty of other trends besides. Scientists have some leeway in deciding which of those patterns deserve attention. Their decisions about which trends to look for could be informed by all sorts of considerations, including everything from an interest in evolutionary progress to an understanding of the limitations of the available fossil evidence. It is possible for different people with different interests to spot different patterns in the same data. Both of the patterns could be very real. Paleontologists only focus on some subset of the patterns and trends that are really there.

Dennett (1991, pp. 35–36) argues that when studying patterns, we often confront a difficult trade-off. On the one hand, it is sometimes possible to offer a simpler, more compact description of a pattern which comes with a very high noise ratio. But sometimes one can also offer a more elaborate, more complicated description of the pattern that comes with a much lower noise ratio. So we get to choose between "an extremely compact pattern description with a high noise ratio or a less compact pattern description with a lower noise ratio" (Dennett 1991, p. 36). In order to illustrate this, Dennett imagines two different people looking at frame E in Figure 6.3. The first, Jones, sees a simple pattern, the bar code (ten rows of ninety dots each, where each row consists of ten black, ten white, etc.), but with a high noise ratio of 33%. But the second person, Brown, takes an interest in the noise, and finds that the noise is not completely random. Brown is able to offer a more complex description of the pattern in frame E which also includes a description of some of the pattern in the noise. Thus, Brown ends up with a much lower noise ratio than Jones does. Another way to capture this difference would be to say that Jones is offering a coarse-grained description of the data, while Brown is offering a finer-grained description. Which is the better way to go? Well, since both descriptions are compressed – i.e., are more efficient than giving the complete 900 pixel bit map – both get at real patterns in the informational sense. Both of the patterns are there; which one we choose to focus on is just a matter of what our interests are.

I think that Dennett's discussion of this trade-off may help to shed some light on one otherwise puzzling feature of the paleontological work on Cope's rule. At some point in the preceding discussion, the following question may have occurred to you: "Paleontologists are most interested in showing us how evolution looks at the largest scales, and in giving us the big picture of the history of life. But the big picture almost seems too easy to paint. At the largest scale, Cope's rule is just obviously true – 3.8 billion years ago, when life first arose, the mean body size of living things was ridiculously small, certainly no larger than the size of a single prokaryotic cell. Today, even though most living things are still microbes, the presence of lots of multicellular plants and animals pushes up the mean body size. So what else is there to say? Of course the mean body size *of all living things* has increased." Earlier on I said that there has obviously been an apex directional trend in body size, or an upward trend in the maximum. At that point you might well have wondered: Isn't it just as obvious that there has been an upward trend in the mean? If you look at it this way, it might just seem silly for Gould or anyone else to question the reality of Cope's rule. It might also seem pointless for paleontologists to do any further empirical research on the subject, since we already know that Cope's rule is real. But the real issue is whether we should look at it this way. It's true that we can say that the mean body size of all living things has increased over the last 3.8 billion years. That is an extremely compact description of a staggeringly large amount of data, and it comes with a really high noise ratio. Paleontologists are not content to mention the obvious largest scale pattern and leave it at that. They want to study the noise, and to see if there are other equally real, but more complex patterns to be described and documented. The main reason for this – as we'll see in the next chapter – is that they want to get at the causes of the patterns.

In setting up his example (Figure 6.3), Dennett includes a description about the underlying process by which frames A through F were generated. They were all made by a computer program that creates the bar code and then goes back and inserts a specific amount of random noise. As we continue to think about the paleontological study of patterns, we need to imagine what it would be like to look at Dennett's frames without knowing anything about the computer program that produced them. Would it be possible to draw any inferences from the patterns concerning the underlying processes that generated them? That depends. Dennett points out, for instance, that if he hadn't told us that frame F is generated in the same way as the others, we might not be able to guess that. Frame F could just as easily have been produced by a program that prints 900 pixels in a 10 by 90 grid at random. In the case of frame F, at least, hypotheses about the underlying processes are, as philosophers of science like to say, *underdetermined* by what we can observe of the frame. Most of the discussion here is focused on the metaphysics of trends and patterns, but the problem of underdetermination is an epistemological one. How much can we know about the underlying processes that generate patterns? That's the big question of Chapter 7.

Trends and evolutionary progress

Why would Gould be so keen to argue that Cope's rule is merely a psychological artifact? Gould was a consistent skeptic about the idea of evolutionary progress, and throughout his career he found a number of different ways of attacking that idea. For him, attacking the idea that Cope's rule is a real trend was just one way of attacking the idea that evolution is progressive.

Sober (1994) offers a helpful way of thinking about evolutionary progress when he writes that "progress = directional change + values" (compare also Ayala 1988). Indeed, this is a helpful way of thinking about claims about progress in any domain. The concept of progress is what philosophers sometimes call a thick concept, because it includes both *normative* and *descriptive* components. In general, a descriptive claim is a claim about how things are, or about what *is* the case, whereas a normative or evaluative claim says something about how things *ought* to be. This is the notorious is/ought distinction that causes so much difficulty for philosophers who work in the areas of ethics and meta-ethics. One piece of advice for anyone interested in taking up the issue of evolutionary progress is simply to respect this distinction between the descriptive and the normative. Paleontology has a special contribution to make to discussions of evolutionary progress by supplying the empirical, or descriptive, component. Paleontologists document the real historical trends and thereby provide us with an occasion to think about whether those directional changes constitute any kind of improvement. When we make any judgments at all concerning whether a given directional trend amounts to progress, we inevitably bring some normative standards to bear.

The following sort of reflection leads me in some moods to think that evolution is progressive, and that it would be crazy or misanthropic to think otherwise. Imagine what it would be like to compare two worlds side-by-side. One is the world as it existed 3.5 billion years ago, when the most complex form of life on the planet was a single-celled prokaryote. The other is the world as it exists today, with gazillions of single-celled prokaryotes as well as a staggering diversity of more complex forms of life, including humans. Which of these two worlds is better, or more valuable, or superior, or more advanced? Surely there is something better about the second world. And if you say that, then you must, at some level, believe in evolutionary progress. Consider the alternative. Suppose you say that the first of the two worlds is better. Or suppose you refuse to make the call, which seems tantamount to saying that the two worlds strike you as equally good, equally valuable. In that case, at the very least, you are denying that the presence of human beings adds any value to the world, or that a world that contains human beings is any better than one that does not. That view seems distasteful.

Many philosophers argue that belief in evolution is perfectly compatible with belief in God, a supernatural person who created the natural world (see especially Ruse 2004). Perhaps the most straightforward way of combining belief in evolution with theism is to take on board everything that scientists tell us about evolution and the history of life, and then just add that God had his reasons for creating a world in which things unfold in just this way. Perhaps God is interested in obtaining a certain outcome by means of evolutionary processes. This is one of several intellectually respectable views about the relationship between science and religion. It does seem to commit one, however, to belief in evolutionary progress. If you sympathize with this theistic view that evolution is a manifestation of God's creative activity, then surely you will say that the second world is, in some sense, better than the first.

In other more misanthropic moods, I find that I can talk myself into denying that there has been any evolutionary progress. One way to induce such moods (for me, personally) is to think about the grandeur, the diversity, and the beauty of past forms of life. The dinosaurs, for example, evolved during the Triassic period, some 225–250 million years ago. They flourished and diversified for many millions of years, even surviving a couple of mass extinction events. When I look at dinosaur skeletons or ammonite fossils from the Cretaceous, they seem to me to be the remains of creatures that were just as beautiful and as fascinating as anything alive today. When I consider the world as it was, say 80 or 90 million years ago, it no longer seems quite so obvious to me that the world as it is today is better or more valuable. I'm inclined to say that the two are just different. Then it occurs to me that by means of a series of reflections of this kind, I could easily get myself to the point of thinking that the prokaryotic world of 3.5 billion years ago is "just different" from ours, but no worse or less valuable. If our species were to go extinct at some point in the next few hundred thousand years, then from a cosmic perspective the short-lived human adventure on Earth might look like a minor offshoot of the much larger tree of life, rather than any sort of culmination of evolutionary history.

I don't want to defend either of these two perspectives here; I do, however, think that it is important in any discussion of evolutionary progress to be able to feel the pull of each view. Moreover, it's important to realize that both of these views involve normative judgments about evolutionary history.

Where paleontology meets meta-ethics

Every judgment concerning progress involves an application of some evaluative or normative standards. What standards should we apply, and where might those standards come from? One suggestion is that natural science itself might supply the standards. Philosopher of biology Timothy Shanahan goes this route, and argues that a reasonable standard of progress might be "the ability to survive and reproduce despite changing environments" (2004, p. 237). This standard is closely connected to the very idea of natural selection, since selection is just a matter of differential survival and reproductive success in a given environment. Notice, however, that Shanahan emphasizes changing environments. Being able to survive and reproduce successfully in the face of dramatic environmental changes is, on this view, what makes some forms of life superior to others. Evolutionary progress occurs as organisms gain more and more independence from environmental conditions.

One concern about Shanahan's proposal is that it might lead to some counterintuitive judgments. To begin with, it is difficult to know how to assess our own ability to survive and reproduce in changing environments. It's true that human beings have shown a remarkable ability to adjust to life in radically different environments, from the arctic to the tropics. However, it might be wise to adopt a wait-and-see attitude with respect to our ability to cope with large-scale environmental changes that we ourselves are causing, such as climate change and biodiversity loss. Intuitively, it seems like we might want whatever standard of progress we adopt to yield the result that human societies represent the "highest" or most advanced products of evolution to date, but it is not entirely clear that Shanahan's standard would yield that result. The prize for coping best with changing environments might have to go to seemingly humbler organisms, such as ferns, that have a strong track record of hanging on through mass extinction events.

Shanahan's attempt to identify a standard of progress internal to evolutionary theory may also be vulnerable to the famous open question argument, due to G.E. Moore (1903/2004). Moore took issue with attempts to define the property of being good in terms of some other natural properties. Imagine, for example, that I declare that "whatever makes me happy, is good." If you were to ask me what the word "good" means, I would just say, "whatever makes me happy." According to my pet theory, if we know that X makes me happy, then it follows *by definition* that X is good. For example, suppose that natural history museums make me happy. From that it must follow, by definition, that natural history museums are good. Moore then argues in the following fashion: granting that museums make me happy, we may still ask "with significance," as he puts it, whether museums are good. That is, given that museums make me happy, it is still an open question whether they are good. The fact that the question, "Are natural history museums good?" still makes sense to us shows, according to Moore, that it can't really be true *by definition* that they are good, given that they make me happy. Now the example I just gave is a silly one; obviously no serious ethicist would ever try to define goodness as what makes her happy. Moore, however, thought that this style of argument could be deployed against any attempt whatsoever to define goodness in terms of some natural property or properties. Those who go in for such definitions commit what he famously called "the naturalistic fallacy." Moore was careful to allow that goodness might be co-extensive with some natural predicates. For example, every organism that has a heart has at least one kidney. Those two predicates – having a heart and having at least one kidney – are coextensive. But it does not follow that the two predicates mean the same thing. They refer to all the same things, but the predicates have different meanings. Normative and non-normative predicates can be co-extensive in the same way, but that does not mean that we can define the normative predicates in terms of the non-normative ones. Moore's argument poses some interpretive challenges that we need not get into here. Many philosophers ignore the details and take Moore to have shown that it is a mistake to try to derive an "ought" from an "is" – a point that many also attribute to the eighteenth-century Scottish philosopher, David Hume.

Moore's open question argument poses a problem for anyone who looks to find a standard of evolutionary progress within evolutionary theory. The only difference is that judgments about progress are comparative judgments to the effect that a later set of conditions is better than an earlier one. Any attempt to define evolutionary progress in terms of natural relations will run afoul of Moore's argument. Suppose, for instance, that we take Shanahan's proposed standard as a definition of evolutionary progress. On such a proposal, what it means to say that evolution has progressed is simply that later organisms are better able than their predecessors to cope with changing environments. Next, suppose we are told that some later species B is better able to survive and reproduce in changing environments than some predecessor species A. We can still ask, with significance, whether B is more advanced than, or better than, A. Yet although it is important to be aware of Moore's open question argument, it is easy to make too much of the argument. Respecting the descriptive/normative distinction does not mean that we have to refrain from making normative judgments about evolutionary history, or about historical trends more generally. That would be too much to ask.

To take an example from ordinary life, think of the process of paying off one's student loans. That involves (hopefully) a directional trend: decreasing amount of principal owed. In saying that this amounts to progress, one is making a normative judgment and implicitly applying some normative standard for determining what counts as better or worse. In a familiar case like this one, it would be odd to construe the claim as an attempt to define the normative relations (better than/worse than) in terms of the natural relations (smaller debt/bigger debt). That would be reading too much into the claim. Instead, all that's going on here is that you have a goal – a condition that you desire to bring about – and you are using that goal as an external standard for assessing progress. It is only because you care about reducing your debt to zero that the historical trend in this case amounts to progress. One plausible suggestion is that all judgments of progress are implicitly anchored to human goals in this fashion.

The foregoing considerations suggest another way of reading Shanahan's proposal. In proposing that "ability to survive and reproduce despite changing environments" serve as our standard of evolutionary progress, we might read him simply as proposing that this is something we ought to care about, or something we ought to value. Nor is this such an unreasonable suggestion. Given that we ourselves live in environments that are changing, we might indeed think that it is a good thing to be able to cope with changing environments. Thus, it might be unfair to read Shanahan as trying to define normative relations in terms of natural ones, or as trying to derive normative claims from descriptive claims alone. Rather, he might be proposing that we look at evolutionary trends with our own distinctively human values and goals in mind. On this view, which does respect the descriptive/normative distinction, we let paleontologists document and describe the main trends in evolutionary history, and then we make judgments about those trends in a way that is informed by our own concerns, preferences, commitments, and goals. On this view, we might also decide that complexity increase represents a kind of progress. Or we might not. Whether we make such a judgment depends on our own goals and values. We need not be shy about making normative judgments concerning evolutionary history, but we do need to be thoughtful about it, because such judgments reflect our own values.

One reason why many scientists remain cautious about the idea of evolutionary progress – and indeed one reason why Gould, in particular, insisted that there is no such thing – is that evolutionary theory does not itself involve any claims about overarching goals or ends. Evolution is not itself a teleological process; it is not, so far as we know, and leaving aside any theologically motivated add-ons, occurring for the sake of anything in particular. The worry is that any talk about evolutionary progress at all will involve a distortion of modern evolutionary theory, or a reading of certain human goals back into the evolutionary process when evolution is better thought of as a mindless, non-goal-directed process. This is a legitimate worry: Evolution has no goals, and although we might decide that we like complexity (for instance) it would be an error to suppose that complexity is the *telos* or goal of evolution. Nevertheless, it's possible to respect this point without refraining from talking about evolutionary progress altogether. We can make normative judgments about evolutionary history that are informed by our own human goals and values without making the mistake of thinking that our own goals are also nature's.

The great paleontologist, George Gaylord Simpson (1949), held a view similar to the one I've just sketched. Simpson strongly resisted any attempt to say that evolutionary change is progressive by definition. Instead, he argued that:

> Progress can be identified and studied in the history of life only if we first postulate a criterion of progress or can find such a criterion in the history itself. (1949, p. 241)

Any attempt to find a criterion of progress in the history of life itself is liable to run afoul of Moore's open question argument. So the only question is what criterion of progress (if any) we should adopt. Progress is relative to whatever standards we decide to adopt. Simpson worried that our selection of criteria might be self-serving, and that we will naturally give in to the temptation to invoke standards that yield the desired result – namely, that our own species is in some sense better than what came before. Simpson went on to argue that "criteria *not* selected with man as the point of reference still indicate that man stands high on various scales of evolutionary progress" (1949, p. 241). He adopted a pluralistic approach and considered a whole slew of different standards of progress:

- Replacement of one species by another within an adaptive zone
- Increasing fitness/degree of adaptedness relative to a given environment
- Increasing dominance by a particular group of flora or fauna
- Greater ecological specialization
- Greater independence from the environment
- Greater structural complexity
- Greater energy use
- More care for and investment in offspring
- Greater perceptual and cognitive ability

After surveying these standards and arguing that humans come out pretty well with respect to most of them, Simpson claims that "there have been not one but many different sorts of progress" (1949, p. 260). He notes that evolution does not always lead to progress by any of these standards, though it often does. For example, the evolution of our own species does not seem to have been characterized by any increase in ecological specialization.

Although Simpson's move toward pluralism is, in part, an attempt to avoid the pitfall of anthropocentrism, he may still be vulnerable to the objection that most of his proposed criteria seem tailored to yield the result that humans represent the pinnacle of evolutionary progress. Think of all the other criteria one could have adopted. One alternative suggestion is that the more longevity a species has, the better. By that standard, any trend toward longer species duration would count as evolutionary progress. Reef corals and foraminifera have average species durations in excess of 20 million years, whereas mammals have average durations of 1–2 million years (Kemp 1999, p. 178). Our own species comes off looking mediocre at best by this "longer lasting is better" standard. Thus, Simpson's pluralistic list of criteria may also be biased in our own favor. But then again, why is that such a problem? Though he took pains to avoid simplistic anthropocentrism, Simpson also wrote that:

> it is merely stupid for a man to apologize for being a man or to feel, as with a sense of original sin, that an anthropocentric viewpoint in science or in other fields of thought is obviously wrong. (1949, p. 241)

Simpson pointed out that if we adopted a criterion of progress in order to get the result that humans are not higher or better than any other forms of life, that is really no different than adopting a set of standards that stacks the deck in our favor.

There are two different ways of rejecting the idea of evolutionary progress. First, one could show that according to a given standard or criterion, progress has not occurred. Someone who adopts that first approach is still making a normative claim about history. Second, one might eschew talk of progress altogether. One could simply refuse to use that thick concept at all. Perhaps many scientists today feel the temptation to go that second route. According to this second view, which is starkly opposed to Simpson's, talk of progress simply has no legitimate place in natural science at all. Scientists should do their best to compartmentalize things. Although they might want to employ thick normative concepts in the privacy of their own homes, so to speak, they should eschew talk of progress when they are on the job.

Simpson, for his part, doubted whether such compartmentalization is possible. "It is impossible," he insisted, "to think in terms of history without thinking in terms of progress" (1949, p. 139). Here it is important to bear in mind that even those who deny progress may well be "thinking in terms of progress." Simpson's point, I take it, is that it is impossible for us to think about history in a way that is entirely divorced from normative notions. Although he makes no clear argument for this claim, it may well follow from the fact that the history of life on Earth is our own history. How could we think of our own history in non-normative terms? This Simpsonian view that normative judgments about history are unavoidable gets some further support, perhaps, from Michael Ruse's (1996) discussion of the history of the concept of progress in evolutionary thought. Ruse argues that at every step of the way, scientists' views about progress have been influenced by attitudes and values in the broader culture. Indeed, Ruse seems to share Simpson's view that we cannot help but think of the history of life in normative terms:

> In the case of evolution, you cannot have a theory about it without some theorizer, which just so happens to be one of us humans. But we ourselves are part of the evolutionary process ... Hence, there is bound to be a tendency to judge the process from our own perspective. We cannot do otherwise. What this means is that we are forced to value things like the evolution of intelligence, since we are the people asking the questions. (Ruse 1996, p. 537)

Are we really forced to value things like intelligence? I can sometimes talk myself into thinking that intelligence has no special importance in the broader scheme of things, and that a world where the most intelligent creatures are dinosaurs could be just as beautiful and valuable as our own world. But such moods do not usually last very long. At any rate, Ruse's main point is a more general one: we cannot think about evolutionary history at all, except in the light of things we do value and care about – whatever those things might be. It may not be possible to set aside all normative concerns when thinking about directional historical change.

7   Dynamics of evolutionary trends

Paleontology is not simply about documenting patterns and trends in the fossil record. Paleontologists seek to go beyond the surface patterns in order to draw conclusions about the underlying evolutionary processes. If the patterns told us nothing about the underlying processes, then it would be impossible to sustain the claim that studying the fossil record can teach us anything about how evolution works. In this chapter I will continue to use Cope's rule of size increase as a case study. I will try to highlight some of paleontology's recent successes in teasing out the causes of evolutionary size increase, while at the same time raising some skeptical doubts and concerns. While it is sometimes possible to draw interesting conclusions about the causes, or underlying dynamics, of evolutionary trends, epistemological modesty is called for. There is only so much that the fossils can reveal.

Varieties of evolutionary trends

The first step is to draw some distinctions among different kinds of historical trends. Paleontologists usually distinguish *within-lineage* trends from *among-lineage trends* (McShea 1998; Alroy 2000). A within-lineage trend is one that results primarily from evolutionary forces – especially natural selection – that do their work within evolving populations. An among-lineage trend is one that results primarily from species sorting, or the differential extinction and speciation of lineages. Here it's important to recall that species sorting does not imply species selection.

This way of putting things makes it sound as if every evolutionary trend can, in principle, be classified as either within-lineage or among-lineage, but unfortunately things are not quite so simple. Some evolutionary trends could well result from a combination of species sorting and, say, natural selection. It could turn out, for instance, that natural selection favors larger-bodied organisms, thus driving body size increase within lineages, while at the same time, smaller-bodied lineages have a higher extinction risk. This means that the distinction between among-lineage and within-lineage trends is not exclusive. A single trend could fall under both headings at once, and in such cases, the real challenge for scientists is to try to determine the relative significance of the forces operating at the micro- versus the macro-levels. On the other hand, the distinction is exhaustive, even if not exclusive. There is no such thing as an evolutionary trend that is neither within-lineage nor among-lineage. Every trend has to result from something, and the only options are species sorting or else the usual microevolutionary processes.

The idea that species sorting can give rise to evolutionary trends is, as we have seen, one of the inspired ideas of the new evolutionary paleobiology. It is the springboard for the theory of species selection. In light of that, one might expect to find paleontologists carrying out studies in which they first identify a trend and then apply some test to determine whether the trend is within-lineage or among-lineage – or more precisely, to determine the relative importance of the causal contributions from species sorting *vs.* evolutionary forces operating within populations. Species selection would receive a huge boost if scientists could show that some trends are definitely due to species sorting. Unfortunately, this is extremely difficult to do. Alroy (2000) points out that in order to carry out this type of study, paleontologists would have to do two very difficult things. First, they would have to study the extinction and speciation rates of the different lineages in the clade. Sticking with the example of body size evolution, they would have to determine whether those extinction and speciation rates have any relationship to body size. In addition, the scientists would have to determine whether there are any directional changes in body size within lineages in that very same clade. In effect, one would have to do two different studies and then put the results together. Perhaps because of the sheer difficulty of such a project, paleontologists have tended to focus more on the distinction between passive and driven trends, which we first encountered in Chapter 6 in connection with Gould's attack on Cope's rule (Alroy 2000, p. 320).

In general, a trend or pattern can be *biased* or *unbiased*. This distinction is easiest to think about with respect to a trait, such as body size, that varies continuously. A trend toward larger body size is biased when the probability of size increase is greater than the probability of size decrease. Where size increase and decrease are equally probable, the trend is unbiased. In a

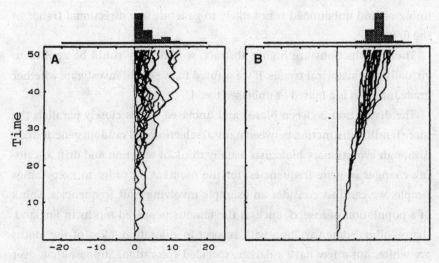

Figure 7.1 Passive *vs.* driven trends, from McShea (1994). These two diagrams show the outputs from two runs of a computer simulation. Trend A is passive, while trend B is driven. Reprinted with permission from Wiley Blackwell publishers.

computer simulation such as the MBL model, one can represent body size as increasing or decreasing by a certain amount with each "turn" of the simulation. In order to keep things simple, it helps to assume that the amount of increase or decrease per time interval is always the same – say, one unit per turn. We can introduce complications by allowing variation in the size of the increases or decreases. Happily, though, those further complications do not affect the basic distinction between biased and unbiased trends. A trend is *bounded* when there is a fixed upper or lower limit to the value of the trait in question. For example, body size increase would be a bounded trend if there were (as indeed there probably is) a fixed minimum or maximum body size for creatures of a certain type.

Paleontologists do not usually talk about biased or bounded trends as I have introduced those concepts here. Rather, the more familiar distinction is between *passive* and *driven* trends (see Figure 7.1). I have in effect broken that distinction up into two sub-distinctions. Any biased trend is driven. The rough idea is that there is something – whatever it is that happens to be responsible for the bias – pushing or driving evolutionary change in a certain direction. A *passive* trend, by contrast, is a trend that is both unbiased and bounded. A driven trend can be either bounded or unbounded. A system that is both unbiased and unbounded is not likely to generate any directional trends in the first place.

These distinctions are highly abstract, so that they could be applied to virtually any historical trends. If we wanted to, we could investigate whether grade inflation is a biased or unbiased trend.

The distinction between biased and unbiased trends closely parallels the more familiar distinction between natural selection and random genetic drift. Although evolutionary biologists usually think of selection and drift as caus-ing changes in gene frequencies, for the moment, in order to keep things simple, we can just consider an example involving trait frequencies. Think of a population of insects, such as the famous peppered moths in England, that exhibits some variation with respect to coloration. Most of the moths are white, but a few have a darker, speckled coloration. Suppose that, over time, the frequency of the speckled moths increases – a clear example of a directional trend. We might reasonably ask whether that trend is biased or unbiased. With each generation, is the probability of an increase in the frequency of individuals with darker coloration greater than the probabil-ity of a decrease? If so, something must account for this bias, and the natural candidate is natural selection: individuals with the darker, speckled coloration must somehow have an edge in the Darwinian struggle for existence. (In the real story, soot from the factories in England was covering the birch trees and turning them brown.) If the trend occurs in the absence of a bias, then it's tempting to say that the trend is due to random genetic drift. There could be other factors involved, such as migration, but drift is the leading candidate. Here as elsewhere, the concepts that paleontologists use when thinking about macroevolution closely resemble those that population biologists have long used in thinking about microevolution (compare Millstein 2000, p. 622).

Dan McShea (1994) has done more than anyone else to clarify and standard-ize the distinction between passive and driven trends. In presenting those dis-tinctions up to this point, I have framed them in probabilistic terms. McShea, however, suggests that one can just as easily think of those distinctions in terms of forces acting on an evolving clade (Figure 7.2). In a driven trend, the forces acting on the clade are relatively homogeneous. They act more or less uniformly to "push" the clade in a certain direction through the state space. The term "state space" here just refers to the range of possible values for the trait under scrutiny – in this case, the range of possible body sizes. A passive trend, by contrast, is associated with heterogenous forces. On the one hand,

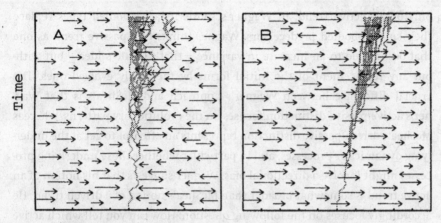

Figure 7.2 Passive *vs.* driven trends, understood in terms of evolutionary forces, from McShea (1994). The driven trend B is the result of homogeneous evolutionary forces that "push" the trend in a certain direction in the state space. In the passive system, A, homogeneous forces maintain a fixed lower boundary, but the rest of the forces acting on the clade are heterogeneous and exert no net effect. Reprinted with permission from Wiley Blackwell publishers.

evolutionary forces act more or less uniformly to maintain the boundary – say, the lower limit on possible body sizes. But aside from that, the forces acting on the clade are mixed and do not "push" it in any particular direction. Most paleontologists switch back and forth freely between these two ways of talking about the passive/driven distinction, regarding the two approaches as interchangeable, and I will follow that practice here.

Thanks to Gould (1988b), many scientists also think of passive trends as involving increases in the total variance of a clade. Gould imagines a clade starting out at or near a fixed lower boundary for body size. Over time, the clade gets more and more diverse, with an increasing variety of different body sizes in different lineages. Since the whole process started out near the minimum possible body size, there is, in a sense, nowhere to go but up. As variance increases, the mean body size of the clade will increase as well. Passive trends that involve no directional bias typically do involve increases in variance.

Not all paleontologists agree about the basic concepts. Although I will follow McShea's usage here, Wagner (1996) proposes a distinction between active and passive trends that is close to, but not quite synonymous with, McShea's driven/passive distinction. Wagner defines an active trend as one in which "the derived morphologies replace the initial morphologies over time" (1996, p. 992). The key idea here is that of replacement. Body size increase would only be an active trend, in Wagner's sense, if larger-bodied types replace their smaller-bodied predecessors. Wagner thinks of a passive trend as one that occurs due to an increase in variance – that's Gould's idea – but without any replacement of the initial forms by the newly evolved ones. The crucial difference between Wagner's approach and McShea's is that Wagner's active/passive distinction focuses on the evolutionary patterns, whereas McShea's driven/passive distinction represents an attempt to get at the underlying dynamics or processes, and in particular whether those underlying processes might involve a directional bias. Wagner suggests that his notion of an active trend is somewhat broader than McShea's notion of a driven trend. He accordingly focuses on the following question: How can you tell which active trends are also driven? We may as well cut to the chase and stick with the question of how to tell whether a trend is passive *vs.* driven in McShea's sense of those terms.

Earlier we saw that the distinction between within-lineage and among-lineage trends is not exclusive, because a trend could result partly from forces operating within populations, and partly from species sorting. This might make one wonder whether the passive/driven distinction is exclusive. Could a trend be partly passive, partly driven? Wang (2001) thinks so, and he introduces some statistical techniques designed to "quantitatively decompose a trend into two components: a driven portion and a passive portion" (2001, p. 849). Indeed, Wang thinks that passive and driven trends "represent extremes of a continuum" (2001, p. 851). It's possible, I think, to appreciate the conceptual point that Wang is making without getting into the technical details of his approach. In principle, there are two ways in which a trend could be partly passive, partly driven. First, directional biases – say, toward larger body size – could come and go over time. Imagine a clade that is evolving over a period of 10 million years. For the first 5 million years, there is no directional bias toward larger body size, and the clade simply does a random walk away from the lower boundary. But then, 5 million years into the process, something happens to create a directional bias that pushes the clade further toward larger body size. If we look at the trend over the whole 10-million-year interval, it will be complex: partly passive, and partly driven. Notice that this phenomenon is closely related to a point made in Chapter 6 – namely that the reality of trends is subject to scaling effects. If we examined this case at a higher resolution, zooming in on the 5-million-year timescale, we could just say that there are two trends, one passive, and one driven. There is a second way in which a trend could be both passive and driven, as Wang suggests. Suppose scientists observe a trend toward larger size in an entire clade. One part of that clade could be experiencing a strong directional bias toward larger size, but other parts might not. In that case, it would make sense to say that the overall trend has both a passive and a driven component. Again, this is related to a point made in Chapter 6 – namely, that the reality of a trend depends not only on the time interval, but on the clade one chooses to look at. In this case, if we zoomed in and looked at the sub-clades, we might find a driven trend in one, and no trend at all in the other. So Wang, I think, makes a significant conceptual point. However, it is also important not to forget that the distinction between a biased and an unbiased trend is an all or nothing distinction. Once we settle upon looking at a particular clade and a particular time interval, then the question whether a trend in that clade is biased should have a simple yes or no answer.

There is, perhaps, another way in which one could think of the distinction between passive and driven trends as admitting of degrees. Consider a fair coin, with a 50/50 chance of landing heads *vs.* tails when tossed. Next, imagine a coin with heads on both sides, so that the probability of landing heads is 1. That represents the limiting case of a bias toward heads. Those two cases do represent two ends of a continuum. We might similarly think of an unbiased (passive) trend and a perfectly driven trend as extreme ends of a continuum, though this is probably not what Wang has in mind.

Finally, it might seem at first glance that every driven trend must also be a within-lineage trend. That's because natural selection is typically assumed to be doing the driving, and selection is a force that operates within evolving populations. We should take care to leave open the possibility, however, that a trend in the mean body size of some clade might be driven by species sorting. Differential speciation and extinction of smaller- *vs.* larger-bodied lineages could be the source of a directional bias with respect to changes in the mean for the entire clade.

Natural selection *vs.* constraints

When paleontologists identify a real trend in the fossil record, one of the first questions they ask is whether it is passive or driven. If a trend is driven, where does the bias come from? And if it is bounded, what explains the existence of a lower (or upper) limit on body size? One possible explanation

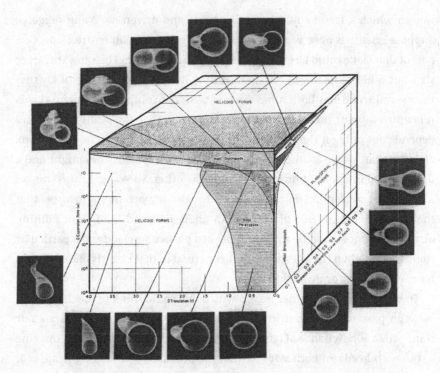

Figure 7.3 D.M. Raup's (1966) three-dimensional morphospace. Shaded areas indicate regions of the morphospace occupied by real biological forms. Reprinted with permission from SEPM (Society for Sedimentary Geology).

concerns natural selection; another possible explanation has to do with biological constraints.

The concept of a constraint is closely related to the concept of biological possibility. In order to think a bit more clearly about constraints, it will help to consider the related concept of a morphospace – roughly what I was calling a "state space" earlier on. One classic example of a morphospace is David Raup's analysis of possible shell forms (see Figure 7.3). Raup's diagram has three dimensions because he looked at three different aspects of the geometry of shell coiling. The shaded areas of this three-dimensional morphospace are (or were) occupied by real organisms. The talk of "space" here is merely metaphorical. Each point in the three-dimensional space represents a different shell shape. One virtue of this kind of approach is that it enables scientists to focus on the unoccupied areas of the morphospace and ask why they are empty. One possibility is that the blank spaces represent morphologies that are somehow less fit, so that natural selection never drove any lineages into those areas. Another possibility is that some other kind of biomechanical or developmental constraint prevented those areas from being occupied. In general, it is helpful to think of a constraint as something that blocks off some region of a morphospace. Even if types falling within that region would be relatively fitter, constraints of various kinds prevent natural selection from driving any lineages into that area. Constraints render some forms biologically impossible.

Before going any further, I must add one caveat. In explaining the distinction between selection and constraints, I have helped myself to the notion of biological possibility, and that notion itself stands in need of further clarification. It is not at all obvious what biological possibility amounts to, or even whether there is such a thing. This is also closely connected to the fraught issue of whether there are any distinctive biological laws, for you might think that something is biologically possible if and only if it conforms to the laws of biology. I propose to leave these difficult issues for another day and to work with this rough-and-ready understanding of biological constraints.

At first glance it looks like the distinction between passive and driven trends is closely related to the distinction between natural selection and constraints. Scientists have long thought that natural selection might typically favor size increases, and if so, that could explain the presence of a directional bias toward larger size. Hone and Benton (2005) offer the following list of potential benefits that might come with larger size:

- Increased defense against predation
- Increase in predation success
- Greater range of acceptable foods
- Increased success in mating and intraspecific competition
- Increased success in interspecific competition
- Extended longevity
- Increased intelligence (with increased brain size)
- At very large size, the potential for thermal inertia (e.g., sauropod dinosaurs and tuna)
- Survival through lean times and resistance to climatic variation and extremes

Indeed, Kingsolver and Pfennig (2004) actually tested the idea that natural selection favors larger-bodied organisms by conducting a review of recent studies of the evolution of birds, and they did find some evidence of selection in favor of larger size. And if natural selection seems like the obvious explanation for the existence of directional biases in evolution, then constraints seem like the equally obvious explanation for the existence of boundaries in the morphospace. If, for example, there is a minimum size for mammals – if the pygmy shrew and the bumblebee bat are about as small as any mammal could possibly be – the likely explanation is that there is some biomechanical constraint at work, something that renders smaller mammals impossible. Thus, it seems that if scientists could only determine whether a trend is passive or driven, they could thereby tell whether the trend is due to natural selection or to biological constraints. And that would be a way of getting at the deeper causes of the trend.

Unfortunately, nothing in macroevolutionary theory is quite that simple. To start with, a lower limit on body size could be due to natural selection just as easily as to any biological constraint. Suppose, for instance, that there is some predator that is unable to kill adult shrews above a certain size. But if an individual shrew happens to be a bit smaller than its peers, it makes an easy dinner for this particular predator. A population of predators could thus be just as effective as biomechanical or developmental constraints at imposing a lower limit on body size. In a case like this, it might even make sense to say that it's just not possible for the shrews to get any smaller, because natural selection is working against that. (This is one reason why the concept of biological possibility is so messy – it is related both to the concept of selection and to the concept of constraint.) Species sorting is another mechanism that could maintain a lower boundary. If small body size comes with heightened extinction risk, that could help explain why we never see mammals smaller than pygmy shrews or bumblebee bats.

Moreover, natural selection is arguably not the only thing that can cause a directional bias in evolution. Dan McShea (2005), inspired by some suggestions in Gould (2002), argues that some large-scale evolutionary trends might even be driven by constraints, rather than by selection. This idea is so fascinating and counterintuitive that it deserves a closer look.

Can constraints drive trends?

McShea (2005) shows how complexity increase could be a constraint-driven trend. Even if he turns out to be right about complexity increase, it is a further question whether other large-scale evolutionary trends, such as size increase, could turn out to be constraint-driven. The main reason for looking at McShea's work here is to drive home the point that "driven" does not necessarily mean "driven by natural selection."

Biological complexity is a notoriously difficult notion to define (see e.g., McShea 1991 for discussion). In his 2005 paper, McShea proposes to understand complexity in purely structural terms as the degree of differentiation of something's parts. One special case of this kind of differentiation of parts is the number of cell types that an organism has, but McShea prefers a more generalized version of this idea. In order to make this idea vivid, imagine two structures built out of Lego blocks. Both contain exactly the same number of parts – say, 100 blocks each. The first structure, however, is made of 100 blocks that are exactly the same, while the second is made from two different kinds of blocks. The second is more complex in the sense that its parts are more differentiated. McShea wants to set all questions about functionality (or in biological contexts, adaptiveness) to one side in order to focus entirely on structure.

The centerpiece of McShea's argument is the so-called *internal variance principle*. He acknowledges that this principle closely resembles an idea defended by the nineteenth-century evolutionist and contemporary of Darwin, Herbert Spencer. Few evolutionary biologists these days take Spencer's work very seriously, and McShea is self-consciously trying to revive a Spencerian approach to understanding large-scale trends, especially complexity increase. According to the internal variance principle, any heritable changes that occur in the parts of an organism will tend to increase the differentiation of the parts. Imagine an organism with three parts, each of which has a certain starting value in some trait m. Now imagine a simulation in which each part has a certain probability of undergoing a mutation at any given turn. Mutations result either in increases or in decreases in the value of m. As this process continues over time, the amount of variance among the three parts with respect to trait m will naturally increase. The parts will grow more and more different from one another over time, thanks to nothing more than randomly occurring mutations (see Figure 7.4). McShea argues that this internal variance principle will create a bias in favor of increasing differentiation among the parts of organisms – and since structural complexity is a matter of differentiation of parts, that amounts to a bias toward increasing complexity. Surprisingly, the directional bias has nothing whatsoever to do with natural selection. Natural selection could, of course, act to maintain the increases in

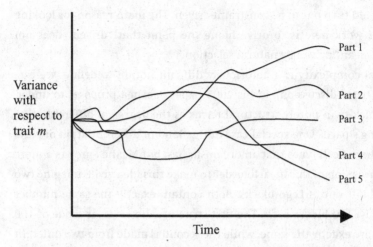

Figure 7.4 The internal variance principle. Parts 1 through 5 all start out with the same value for m. Over time, random mutations (i.e., increases or decreases in m) lead to an increase in the degree of internal variance.

internal differentiation after they arise, but it could just as easily work against them.

Although McShea's argument shows that driven trends need not be driven by natural selection, it may not be quite right to call this a "constraint-driven trend." Whether that term is appropriate depends on what one means by "constraint." If we use "constraint" as a catch-all term to refer to any cause of a boundary or bias in evolution other than selection, then McShea's terminology is perfectly reasonable. But if we use "constraint" more narrowly to refer to a limitation on what is biologically (biomechanically, developmentally) possible, then the internal variance principle may not count as a constraint. A constraint-driven trend could occur if (a) there is some minimum amount of complexity needed in order for organisms of a certain type to be viable, and (b) that minimum shifts upward over time. In the narrower sense of "constraint," a constraint-driven trend would have to involve some change in what is biologically possible, and it isn't entirely clear that McShea's internal variance principle involves such a change. But this is merely a terminological issue. The point is just that driven trends need not be driven by selection.

We should also distinguish between statistical trends and inertial tendencies, such as (in Newtonian physics) the tendency of an object to remain in uniform rectilinear motion unless acted on by something else. Most biologists associate inertial tendencies with outdated notions of orthogenesis, and few think that they have any appropriate role to play in evolutionary theory. But at the same time a driven trend looks a lot like an inertial tendency (Rosenberg and McShea 2007, p. 147). If we say that there is a bias in favor of larger size that is somehow built into the evolutionary process, how is that really any different from saying that there is an inertial tendency toward size increase? The close similarity between driven trends and inertial tendencies is reinforced by the following observation. The fact that there is an inertial tendency toward change in a certain direction does not automatically mean that such a change will occur. For example, populations might have a tendency to grow (and keep growing) at a geometrical rate, as Darwin thought, even if they hardly ever do grow at such a rate. The reason is that other external factors can and do prevent populations from growing exponentially. A process can involve a directional bias without necessarily involving a directional trend. You can flip a weighted coin, with a .6 probability of heads, 10 times and still get a 50/50 distribution of heads vs. tails. A directional bias in evolution, or in any process for that matter, need not generate a persisting directional trend.

In spite of the close affinity, it would be a mistake simply to equate driven trends (or more precisely, directional biases in evolution) with old-fashioned inertial tendencies. In modern evolutionary theory, for the most part, natural selection obviates the need for any inertial tendencies. Selection typically explains the directional biases. Having said that, McShea's internal variance principle really does harken back to an earlier time when inertial tendencies had a home in macroevolutionary theory. McShea is saying, in a Spencerian vein, that complexity (understood as internal variance) has a natural tendency to increase over evolutionary history. Other factors, including natural selection, might act to prevent that increase, but the tendency toward complexity increase is always there. One admirable thing about McShea's work is that he takes some of the mystery out of the notion of an inertial tendency, and shows that the tendency toward increased differentiation of parts follows from a few simple facts about part/whole complexity and mutation.

Is Cope's rule a driven trend?

The distinction between passive and driven trends concerns the underlying causes or mechanisms that generate observable patterns. Or as McShea (1994) more carefully puts it, the distinction is between different kinds of underlying processes. If scientists can determine whether a trend is passive or driven based on a study of the fossil record, that would mean that patterns can reveal something important about processes.

Alroy (1998) set out to determine whether size increase in mammals is a driven trend. He also wanted to determine whether size increase is a within-lineage trend or an among-lineage trend; indeed, his is one of the few studies to tackle both of these questions. Alroy looked at over 15,000 fossil specimens from 1,534 mammal species. Because not every specimen is complete, he measured the size of the lower first molar, and then used those measurements to estimate total body mass. (That is a major advantage of studying body size: there are techniques for estimating body size based only in the size of a single tooth.) Next, Alroy carried out a version of an empirical test that McShea (1994) had suggested a few years earlier. If you want to know whether body size is increasing within lineages over time, a natural approach is to compare ancestor and descendant species. If it turns out that the descendant species are typically bigger than their ancestors, then that is good evidence that the trend is driven. Not only that, but the so-called ancestor/descendant test would also show that the trend is occurring within-lineages, and that it is due to natural selection (or to some combination of microevolutionary forces) as opposed to species sorting.

One problem with the ancestor/descendant test is that it requires a lot of background knowledge of phylogeny: one has to know which species descended from which. Alroy found an ingenious way around this problem. He selected pairs of congeneric species – i.e., species classified as belonging to the same genus. Then he just treated the older of the two species as the "ancestor" and the younger one as the "descendant." This might seem like cheating: the fact that species B is younger than species A, while both belong to the same genus, does not mean that B evolved from A. Still, Alroy reasoned that many of the pairs identified in this manner really would be ancestor/descendant pairs. Furthermore, he argued that this approach is very conservative with respect to the question he wanted to investigate. If anything, this method of identifying pairs would reduce the probability of seeing a genuine within-lineage trend. If a trend appears anyway, that would mean a strong showing in favor of Cope's rule.

Alroy's approach also helps address the question whether Cope's rule is a driven trend. Once the ancestor/descendant pairs have been identified, one can then estimate the average amount of size increase. When Alroy did this, he found that, on average, the younger ("descendant") species were 9.1% larger than the older ("ancestor") species – pretty clear evidence that Cope's rule is a driven, within-lineage trend, at least in mammals. One way to think about Alroy's method is by using the simple analogy of tossing a coin. If you toss a coin twenty times and obtain heads on sixteen out of twenty tosses, that doesn't prove that the coin is unfairly weighted towards heads; it's always possible to toss a fair coin 100 times, or 1,000 times, and get heads every time. The pattern does, however, provide evidence of a bias. That is essentially the argument that Alroy is making about size increase in mammals: if within-lineage size increases and decreases were equally probable, then that 9.1% average size increase would be a real surprise.

Once you know that a trend is biased, one interesting question to pursue is whether the bias has remained constant over time (Turner 2009a). It is not unrealistic to suppose that a directional bias can change in strength. For example, if the bias is due to natural selection, then selection might favor size increase strongly at one time, and in certain environments, but less strongly at another time, relative to different environmental conditions. This would be analogous to carrying out a long series of coin tosses but switching coins along the way, using a coin with a .6 probability of heads for a while, and then swapping it out for a different coin with a .7 probability of heads. Although there may be limits to how far scientists can go in identifying changes in the strength of directional evolutionary biases, Alroy (1998) even took some steps in this direction. He grouped his ancestor/descendant pairings according to the time intervals in which they occurred. This approach made it possible for him to compare the average size increases taking place early in the Cenozoic era with size increases taking place more recently. He found that the average size increase in mammals during the Paleocene, roughly 65–55 million years ago, was only about 2.7%. But the more recent Pleistocene, starting about 1.8 million years ago, saw a much more serious 21% size increase. If anything, it looks like the strength of the bias toward larger size increased during the course of the Cenozoic.

McShea (1994) also proposed a "stable minimum" test for driven trends. The stable minimum test focuses on the body size of the smallest-bodied lineage in the clade. In the case of mammals, the idea is to look at the smallest mammals 65 million years ago, and the smallest mammals of the more recent Pleistocene period, in order to see whether the size of the smallest members of the clade has increased at all. McShea points out that if the minimum remains stable, that counts as evidence that the trend is passive. In a passive trend, where the clade starts out near a fixed boundary in the morphospace and does a random walk away from it, some lineages within the clade can be expected to remain near that boundary and occasionally bump up against it. Therefore, if there is a significant increase in the minimum size, that strongly suggests that the trend is driven, and that some evolutionary forces are pushing all the lineages in the clade away from the lower boundary. Alroy found that the size of the smallest mammals has not changed much. The smallest mammals alive today – the pygmy shrew and the bumblebee bat – are about the same size as the very smallest mammals of earlier times. Does this fact count against the claim that size increase is driven?

One possible explanation is that the directional bias toward larger size was never present, or if present, was never very strong, in the smallest mammal lineages. Just as the strength of the bias can, in principle, vary over time, so it can also vary across lineages. But this, in a way, merely describes the phenomenon in need of explanation. If there was a variation in the strength of the bias toward larger size across different mammal groups, one would want to know more about the causes of the variation.

Alroy, for his part, suggests that the stable minimum can be explained by positing the existence of two size optima for mammals: one relatively small size optimum that was achieved early on in the Cenozoic, and a much larger size optimum (1998, p. 732). A size optimum acts as a kind of "attractor" in the state space, or the morphospace. Just as the earth's gravitational field exerts a force on nearby objects and draws them inward, an optimal area in the state space is an area that selection will drive lineages toward. The existence of the smaller size optimum would help explain the stable minimum size for mammals. Once lineages were comfortably situated near that optimum, natural selection probably would not drive them very far away from it. Alroy hypothesizes that "there may not have been enough time during the Cenozoic for the distribution to expand and envelop the upper optimum," which would explain why size increase shows up as a driven trend (1998, p. 732). One concern about Alroy's argument here is whether it makes sense to talk about an optimal body size range for an entire clade, such as mammals. Mammals live in many different environments and exhibit a huge amount of variety. Intuitively, it seems that optimal size needs to be defined relative to some specific ecological niche or selective regime. What could it mean to say that there is an optimal size range for an entire clade?

To summarize: Alroy has found clear evidence that Cope's rule is a driven, within-lineage trend in mammals, although the stable minimum remains something of a puzzle. From a more philosophical perspective, Alroy's work may seem a little bittersweet. On the one (sweet) hand, it represents a success-ful attempt to infer process from pattern. Alroy is using the rigorous statisti-cal techniques of the new paleobiology (I have spared many of the details) in order to get at the underlying causes of evolutionary change. This suggests that studying the fossil record really can teach us something interesting about how evolution works. On the other (bitter) hand, what the fossils seem to be teach-ing us, in this case, is that natural selection is really important in evolution, and indeed that selection is the main driver of within-lineage size increase in mammals. That is bad news for some paleontologists, such as Gould, who have wanted to downplay the importance of selection. Moreover, Alroy's work suggests that in this case, at least, the large-scale macroevolutionary trend is a mere by-product of evolutionary processes going on within evolving lineages. Although he is careful not to rule out the possibility that species selection has played a role, especially in explaining why there might be a fixed lower bound-ary on mammalian body size, the finding of a within-lineage trend seems like a victory for the view that macroevolution is reducible to microevolution.

Underdetermination

Are there any limits to how much we can know about the causes of large-scale evolutionary trends? That is one kind of question that philosophers of science may helpfully ask as part of their engagement with a particular area of scientific research (Turner 2005; 2007). Underdetermination was a major issue in the scientific realism debate that raged during the 1980s and 1990s, and it remains a problem that philosophers of science continue to think about. In this section I will show one way in which claims about underlying evolutionary processes can be underdetermined by the observable patterns (compare Turner 2009a). The aim here is to provide an illustration of the kind of epistemological problems that evolutionary paleontologists might run into. First, though, it will help to survey some recent work in the philosophy of science.

The concept of underdetermination is closely related to an idea known as the Duhem–Quine thesis, after the French physicist, Pierre Duhem, and the American philosopher, W.V.O. Quine. In his book, *The Aim and Structure of Physi-cal Theory*, first published in French in 1906, Duhem pointed out that scientists typically do not test claims about the world one-by-one when they do observations and experiments. Instead, they test bundles of claims (Duhem 1991). According to one simple understanding of scientific testing, scientists begin with a theory T and derive from it some prediction O about what we should observe under certain conditions. Then they check to see whether O is true. Duhem argued that theories never tell us what we can expect to observe without the help of lots of other claims that do not, strictly speaking, belong to the theories in question. Philosophers have come to call these further claims *auxiliary assumptions*. Some auxiliary assumptions almost seem trivial, but they are nonetheless crucial for deriving the observational consequences of any theory. For example, the theory of the Nemesis star (Chapter 6) says that if astronomers point their telescopes in the right direction, they should be able to observe a star whose pathway through the galaxy has certain properties. (As we saw earlier, no such star has been observed.) But in order to derive that prediction, scientists would have to assume, among other things, that the lenses of their telescopes are clean, that the telescopes are functioning properly, that all stars are detectable through telescopes in the same way, and so on. For another example, consider the way in which Darwin's theory predicts the occurrence of transitional forms in the fossil record (Chapter 2). In order to derive that prediction, one needs to make a host of further assumptions about the fossilization process and the degree of completeness of the fossil record. As we saw earlier, Darwin insulated his theory from empirical criticism by suggesting that some of the assumptions necessary for deriving this prediction might be false – or in other words, that the geological record is massively incomplete.

Duhem showed that when our observations come out wrong, we never know for sure whether the problem lies with the theory being tested, or with the auxiliary assumptions that are necessary for deriving any observational consequences from the theory at all. We have to make a revision somewhere, but because claims about the world are always tested as part of a bundle, we never know for sure which claims in the bundle are mistaken.

Quine (1951) took this idea and ran with it, carrying it to its logical extreme. He defended a radical *holism* about scientific testing. Whereas Duhem suggested that we always test claims in bundles, Quine argued that whenever scientists perform an experimental or observational test, they are in effect testing the whole of science, all at once. The purpose of science as a whole is to help us anticipate future experience, and when experience surprises us – when our observations contradict what our scientific worldview tells us to expect in any given case – we must make some changes to our scientific picture. In doing so, however, we have a great deal of slack, or leeway, in deciding what revisions to make. Quine points out that we are usually disposed to be conservative and to make the smallest changes to our scientific picture that we can get away with. His more general point, however, is that there is nothing about our experiences of the world that dictates what revisions we should make to our total scientific picture. When experience doesn't co-operate, we must make revisions, but how we revise is largely up to us. Quine sometimes made this point by saying that our scientific picture of the world is *underdetermined* by our experience, or by our experimental results and observations.

In general, underdetermination occurs whenever there is slack between our observations, or our evidence, and what our theories tell us about the world. Given a certain set of observations or a certain body of evidence, the world could be this way or that way. The evidence does not discriminate between those two possibilities. Some philosophers of science, such as Quine, have argued that underdetermination is just a fact of life in science. According to this claim, known as the *underdetermination thesis*, scientific theories are generally underdetermined by the evidence. Consequently, where scientists do all believe in a single theory, we cannot really explain why they believe that theory by saying that they were led to it by the evidence. Some philosophers sympathetic to the underdetermination thesis have thought that if you want to explain why scientists accept the theories they do, you have to focus more on sociological and professional considerations. This line of thought is just one of several paths toward a *social constructivist view* of science, which questions the rationality of the scientific enterprise. (For helpful discussion, see Hacking 1999, as well as Parsons 2001.) Although Quine himself did not go this route, some social constructivists have used the underdetermination thesis to help defend their views. Other philosophers have tried to show that the underdetermination thesis is false, at least when it is taken as a general claim about all of science (Laudan and Leplin 1991).

In recent years, philosophers of science have moved away from the general question whether scientific theories are always underdetermined by the evidence in order to focus on more localized epistemological problems (Stanford 2001; Turner 2005). The philosophers who have taken this new approach to underdetermination are less interested in making *a priori* arguments about science in general, such as the argument for the Duhem–Quine thesis, and more interested in looking at the ways in which smaller-scale underdetermination problems make a difference to scientific practice. In general, we can say that a local underdetermination problem occurs in science when the following conditions are met:

1. *The rivalry condition.* Two hypotheses, *H1* and *H2*, are incompatible genuine rivals that afford explanations of observable evidence;
2. *The equal current evidence condition.* There is an evidential tie between *H1* and *H2*, meaning that:
   (2a) the available evidence supports *H1* and *H2* equally well, given current auxiliary assumptions, and
   (2b) there are no other evidential considerations, such as simplicity or coherence with the rest of our knowledge, that can help break the tie between *H1* and *H2*;
3. *The future evidence condition.* Although it's always possible that new evidence will crop up, no one has reason to think that there will be new evidence that helps to break the tie any time soon.

These conditions describe a type of epistemic situation that can arise in science. With this analysis in hand, we can pose some interesting questions about particular areas of scientific research. How often do these underdetermination problems crop up? When they do crop up, what is their source? How do (and how should) scientists deal with them? Before going on to look at an example, I should mention a couple of features of this analysis.

First, it might seem that in treating underdetermination as an evidential tie between two hypotheses, I am ignoring what I said a moment ago about Quine's holistic view of science. If you find Quine's holism appealing, you might want to reformulate the above analysis so that it involves a comparison of two total scientific pictures, which are exactly alike except for one detail: one total scientific picture includes *H1*, while the other includes *H2*. Then you could read this as a description of a case where two total scientific worldviews are underdetermined by the evidence, but those worldviews differ only in what they say about some localized scientific issue.

Second, condition (2b) is deliberately formulated in such a way as to bypass an important philosophical issue. Philosophers disagree about whether simplicity, explanatory power, coherence, and other such theoretical virtues carry any evidential weight. If one theory is simpler than another, does that make it likelier to be true? Condition (2b) could be satisfied if the two hypotheses,

H1 and *H2*, are equally simple. But it could also be satisfied if simplicity were to turn out not to carry any evidential weight at all. (2b) should be read as saying that appeals to simplicity won't help to break the evidential tie, while allowing that there are different reasons why such appeals might not help.

Third, with respect to condition (1), we might want to allow for cases in which conflicting hypotheses are not genuine rivals. For example, one hypothesis is that mammals have always been as big as they are today, but God, wanting to fool us, messed around with the fossil record so as to create the illusion of size increase. This hypothesis obviously conflicts with Alroy's explanation of the pattern in the fossil record. Even worse, this theistic hypothesis can be adjusted so as to ensure that it has exactly the same observable consequences that Alroy's explanation has. The two hypotheses seem to be evidentially tied. Perhaps – just perhaps – one could argue that condition (2b) remains unsatisfied here, and that one of the two is simpler than the other. (Whether that solves the problem, I'll leave to you to think about.) If we cannot somehow rule out the theistic hypothesis, then all bets are off in evolutionary paleontology. Any pattern in the fossil record could have been arranged just so by God, who, after all, is supposed to be all-powerful. We might even give this problem a name and call it *radical pattern/process skepticism*. Many philosophers of science try to avoid this sort of problem by insisting that condition (1) actually fails in this case, since the theistic hypothesis is not a "genuine rival" of the going scientific explanation of the pattern.

I myself have mixed feelings about this move. One needs to avoid radical pattern/process skepticism somehow, and the question is how best to do it. There does seem to be a sense in which the theistic counter-hypothesis is just too easy to come up with. Nor is the theistic counter-hypothesis something that anyone would seriously produce in the course of doing science; rather, it seems imposed on science from the outside. These might be reasons for denying that it is a genuine rival of the scientific explanation. But on the other hand, there might also be good reasons for taking the broader view that any contradictory claims are genuine rivals. What else could it mean to say that two claims are rivals, than that they contradict?

For now, I propose to set the problem of radical pattern/process skepticism to one side. It bears a strong resemblance to the problem of radical skepticism about the past (How do we know that God did not create the world five minutes ago?) and to other radical skeptical problems that keep epistemologists awake at night. To say that philosophers have written a lot about how best to respond to such problems would be a gross underestimation. Even though radical pattern/process skepticism seems like a distinctively paleontological form of skepticism, the question of how best to respond to it actually has very little to do with any details of paleontology. That's a far more general philosophical question that I won't pursue any further here.

Shifting biases, shifting boundaries, and shifting optima

Let's now revisit Alroy's work and take a look at one fascinating detail of his 1998 study. He found that during the early Cenozoic, immediately following the K-T mass extinction event, the directional bias toward larger size in mammals was relatively weak. On average, younger species were only 2.1% larger than the older congeneric species he compared them to. The bias toward larger size shows up much more strongly in more recent times. Indeed, if we could travel back just a short distance in time – say 1.5–2 million years – we'd see all sorts of large mammals, from woolly mammoths to giant ground sloths to glyptodonts. Recent extinctions have wiped out some of the largest terrestrial mammals. It's also important to bear in mind that mammals did not suddenly appear on the scene 65 million years ago; the mammalian fossil record extends way back into the Mesozoic era. Mammals evolved during the Triassic period, some 210–225 million years ago, around the same time that dinosaurs evolved and diversified. That means that mammals were around for a very long time before the fossil record shows any significant trend toward larger size. Alroy's study focused only on mammals of the Cenozoic – the last 65 million years. Why did the mammals remain so small for so long? This fact poses a challenge to the naïve view that natural selection generally favors larger organisms. One possible explanation is that for a long time, there simply was no bias toward larger size. The directional bias first showed up in the early Cenozoic and then grew stronger over time. Another possible explanation is that there had always been a bias toward larger size but that something was imposing an upper limit on mammalian body size. During the Cenozoic, that upper limit was relaxed. Then there is Alroy's own suggestion that there might be several optimal size ranges for mammals, and that for much of their history, most mammal lineages hovered around the smallest of these size optima. It could be that the optimal size range for mammals shifted around in the morphospace, or (and this is more in keeping with Alroy's suggestion) that new optimal size ranges appeared at some point during the early Cenozoic. I've already expressed one concern about that proposal – namely, that it's not clear what it means to say that there is an optimal size range for an entire clade. There is also a worry here about underdetermination: these are three different stories about the dynamics that underlie size increase in mammals. Is there any way to tell which one is true?

Consider another interesting study of Cope's rule (discussed in Turner 2009a). Hunt and Roy (2006) looked at the evolution of a single genus of deep-sea ostracodes, known as *Poseidonamicus* – the "friends of Poseidon." A deep-sea ostracode is a tiny bivalve crustacean, often less than 1 millimeter in length. They are sometimes called "seed shrimp" because if you could peer inside their shells, they would look like miniature crustaceans, with eyes, antennae, and legs. Hunt and Roy studied the evolution of *Poseidonamicus* over the last 40 million years, and they found clear evidence of evolutionary size increase. The earliest populations had an average valve length of 550 microns (about half a millimeter), while the most recent populations had an average valve length of 850 microns. Not only that, but Hunt and Roy were also able to show that *Poseidonamicus* exhibited size increase within lineages – further evidence that Cope's rule is a driven trend (Figure 7.5).

Hunt and Roy also note that there is a generalization in ecology known as Bergmann's rule, which says that variation in body size is correlated with temperature variations. For example, white-tailed deer have a very large geographical range in North America; they flourish from Florida to eastern Canada. The deer living in Quebec are somewhat larger than deer belonging to the same species that live in Florida. The standard adaptationist explanation of this phenomenon concerns surface-to-volume ratio. As body size increases, the ratio of surface area to body volume changes, and it becomes a bit easier for an animal to maintain a constant temperature in a cooler climate. This got Hunt and Roy wondering whether body size evolution in *Poseidonamicus* might be related to ocean temperatures. The standard explanation of Bergmann's rule applies only to warm-blooded creatures, such as mammals; no one really has a good account of why temperature should be related to body size in invertebrates. Nevertheless, Hunt and Roy decided that this relationship is at least worth exploring. It turns out that there is independent geological evidence concerning ancient ocean temperatures. Scientists can take the temperature of the oceans by looking at the chemical content of fossil shells. When marine organisms form their shells, they preferentially use different isotopes of the same chemical element, depending on the temperature. By grinding up

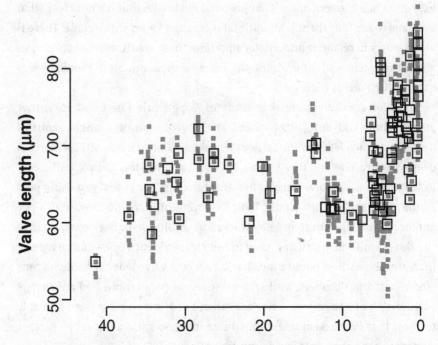

Figure 7.5 Size increase in *Poseidonamicus*, from Hunt and Roy (2006). Each dot represents a single fossil specimen. Each box represents the mean for a collection. Copyright 2006, National Academy of Sciences, USA.

the shells and studying the isotope ratios, it is possible to get a picture of trends in ocean temperatures. When Hunt and Roy plotted the trend in the body size evolution of *Poseidonamicus* against the changes in ocean temperatures, they found a clear but inverse correlation (Figure 7.6). Falling ocean temperatures correlate with increasing size. During the middle of the time interval they studied – from 30 million–10 million years ago – ocean temperatures remained stable, and *Poseidonamicus* stopped getting bigger.

This finding does not prove conclusively that ocean temperature changes are the cause of evolutionary changes in body size, for there could be some other environmental factor which correlates with the temperature change, and which is really the culprit here. It's also important to add that *Poseidonamicus* is just one of many thousands of ostracode genera that scientists have described. But one cannot deny that the finding is highly suggestive.

Nevertheless, there are a couple of different evolutionary stories that one could tell about this case (Turner 2009a). According to one view, the falling ocean temperatures create a bias toward larger size. When the ocean

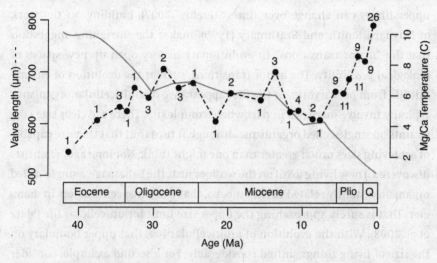

Figure 7.6 Ocean temperature change and size increase in *Poseidonamicus*, from Hunt and Roy (2006). The solid line shows a decrease in ocean temperatures over the last 40 million years, while the dashed line shows the trend toward larger size in *Poseidonamicus*. Copyright 2006, National Academy of Sciences, USA.

temperatures stabilize, the bias disappears. According to a second view, there is a more or less constant bias toward larger size; however, ocean temperatures (somehow) impose a constraint or upper limit on ostracode body size. As temperatures fall, that constraint relaxes, and the longstanding constant bias is able to push the group toward larger sizes. Hunt and Roy's study does not really tell us which of these is the best way to think about how temperature might be related to body size evolution. Lest anyone think that the difference between these two different trend dynamics is merely semantic, consider what the two might say about the relationship between ocean temperatures and natural selection. On the former view – that is, according to the shifting bias story – cooler ocean temperatures create a selection pressure in favor of larger size. According to the shifting boundary story, relatively warmer ocean temperatures create a selection pressure against larger body size, or else they impose a constraint on body size, even though larger size could well have other fitness advantages. These two evolutionary stories – the shifting boundary story and the shifting bias story – say different things about the causes of evolutionary change.

In general, it's plausible to think that there can be upper limits on body size, just as there can be lower limits. It's also plausible to think that those upper limits can change over time. Sterelny (2007), building on the work of Maynard Smith and Szathmáry (1995), makes the intriguing suggestion that the "major transitions" in evolutionary history open up new spaces of biological possibility. The major transitions, such as the evolution of eukaryotic life from prokaryotes, or the first appearance of multicellular organisms, typically involve increases in part/whole complexity. There are clear size constraints on single-celled organisms, although it turns out that they are capable of achieving sizes much greater than one might think. Not long ago, scientists discovered a new living fossil on the seafloor near the Bahamas – a single-celled organism distantly related to the amoeba, that was three centimeters in diameter. That is surely approaching the upper size limit for unicellular life (Matz et al. 2008). With the evolution of multicellular life, that upper boundary on the size of living things shifted considerably. For a second example, consider the argument that paleontologist Peter Ward makes in his recent book, *Out of Thin Air* (2006). Ward thinks that many of the major events in evolutionary history resulted from changes in the amount of oxygen in the atmosphere and the oceans. The fossil record suggests that during the Carboniferous period, insects attained much larger sizes than they do today. The dragonfly *Meganeura* had a wingspan of 75 cm, or about 2.5 feet. Insect body size is limited because of the requirements of respiration. Insect respiratory systems rely heavily on the passive diffusion of air through a system of tiny tracheal tubes that run throughout the insect's body. The bigger the insect body, the less efficient the respiratory system, because it becomes more difficult for the air to diffuse through the tracheal tube system. During the Carboniferous, however, there was a lot more oxygen in the atmosphere than there is today. With more oxygen available in the air, it became possible for larger-bodied insects to have respiratory systems that worked well enough while still relying on passive diffusion. Although this is more speculative, Ward also argues that changes in the amount of oxygen in the atmosphere during the Paleozoic also correlate with changes in the body sizes of the vertebrate land animals living at that time (2006, p. 131). These examples and others like them make plausible the idea that there can be upper limits on body size, and that those limits can change.

Even if we know that Cope's rule is a driven trend, it may not in general be possible to tell the difference between an increase in the strength of a directional evolutionary bias and a shifting of the upper boundary. This is an example of pattern/process underdetermination, where different underlying trend dynamics can generate exactly the same observable pattern. What exactly happened to mammals during the Cenozoic? Was there a new selection pressure in favor of larger size? Or could it be that selection had always favored larger size, but that during the Cenozoic an upper boundary on mammalian body size was relaxed? Both of these stories fit Alroy's data well. Both also raise new questions. If a new bias in favor of larger body size suddenly appeared in the Cenozoic, why? What made larger body size more advantageous than it had been previously? And if an upper limit on body size was relaxed, why did that happen? There is no available evidence that can discriminate between the shifting bias and the shifting boundary stories.

Philosophers of science have disagreed about how widespread local underdetermination problems like this one might be. Carol Cleland (2001, 2002) takes a relatively optimistic view, and argues that, in general, historical scientists can be confident that there is evidence out there that can help answer our questions about the past, although we might not know how to interpret it, and might need new technology to be able to access it. Others, including me in some earlier writings (Turner 2005, 2007; Kleinhans et al. 2005), have taken a more pessimistic stance, and argued that the incompleteness of the fossil record means that underdetermination problems will abound. The wisest approach, perhaps, is to split the difference between the optimists and the pessimists and insist that how much we can know about the past – a question that looks like one for normative epistemologists – is really an empirical scientific question. When we ask how much we can know about the past, we are in effect asking how much information about the past the fossil record actually contains. Here I have described a case where the fossil record has so far failed to answer an interesting question about the dynamics of evolutionary trends. But I should stress, in conclusion, that this particular underdetermination problem only occurs at all because of the rich set of concepts, inspired by computer modeling, that paleontologists have developed. It's only because of their theoretical concepts – especially the notions of biased and bounded trends – that paleontologists can ask questions that the fossil record can't answer. This is a case where an empirical roadblock may also signal theoretical progress.

8   Evolutionary contingency

"[I]f you wish to understand patterns of long historical sequences, pray for randomness"

– Stephen Jay Gould (1993a, p. 397)

Richard Lenski and his research team at Michigan State University have designed an extended experiment in evolution (Lenski and Travisano 1994). Evolution is often difficult to observe in action because it takes so long. It takes many generations for evolutionary changes to accumulate in a population, and so we seldom see such changes happening before our eyes in populations with long generation times. However, Lenski realized that it might be possible to test ideas about medium-scale evolutionary processes by using populations of bacteria. Since bacteria reproduce asexually, scientists can create multiple populations that all have the same genes, with each population living in its own petri dish. Lenski and colleagues created twelve such populations of E. coli bacteria and let the populations evolve in the lab. Once per day, they would add sugars to the dishes for the bacteria to feed on, and the populations would subsequently grow. As the food disappeared, the classic Darwinian "crunch" would occur, and many of the bacteria would die off. The scientists repeated this daily routine for thousands of bacterial generations, a long enough time to observe evolution in action and still not have to wait a ridiculously long time before they could obtain results. They also developed another trick: they stuck some bacteria from the starting populations in the freezer. Later on in the process, they thawed out these ancestors and added them to the petri dishes with their descendants. By that time, the descendants had undergone considerable evolutionary changes. The scientists then let the descendants compete with their own ancestors in the Darwinian struggle for existence. By looking at the relative reproductive success of the descendants *vs.* their ancestors, Lenski and his group found that they could measure the degree to which fitness had increased relative to that particular sugary environment.

Lenski's elegant experiment was inspired by questions that first arose in paleontology. In his classic book on the fossils of the Burgess Shale, *Wonderful Life* (1989), Stephen Jay Gould had argued that evolutionary history is contingent. Lenski and his team hoped to test that claim in the laboratory.

My aim in this chapter is to introduce the debate about evolutionary contingency that Gould initiated. The debate is complex, and it reaches beyond paleontology into philosophical and even theological territory. I begin by presenting the view of evolutionary history that Gould was arguing against.

Visions of inevitable outcomes

In the late nineteenth century, the American pragmatist philosopher C.S. Peirce defended a bold vision of science. He was struck by examples of independent discoveries in science, such as Darwin's and Wallace's discovery of natural selection, as well as Leibniz's and Newton's independent inventions of calculus. He saw that scientists in his own time who used a variety of methods for estimating the speed of light had begun to converge on the same results. Peirce argued that this inevitable convergence is the hallmark of science: if you give scientists enough time, and if they apply scientific methods with enough patience and rigor, then eventually they will always converge on the same results. Peirce even went so far as to use the religious language of predestination to describe this phenomenon. (See the classic essays, "The Fixation of Belief" and "How to Make Our Ideas Clear," in Peirce 1955.) Some scientists have argued that much the same is true of evolution. Given enough time, evolutionary processes will always end up at roughly the same place.

According to Peirce's vision, the eventual outcome of scientific research is insensitive to variations in initial conditions. In other words, it makes no difference where scientists begin; they will always, inevitably, end up at the same place as long as the scientific method is properly applied. Sober (1988) uses a simple model to help illustrate this idea. Imagine that you are standing on the rim of a giant bowl or crater holding a basketball. You can walk around the rim and release the basketball from any point you choose. When you release the ball, it rolls down the side of the crater, across the bottom, and then back up the other side, before rolling back down again. Eventually it comes to rest at the bottom. And it always comes to rest at exactly the same place no matter where along the rim you release it.

Striking examples of convergent evolution seem to make plausible the idea that evolutionary processes might be insensitive to initial conditions. Some examples of evolutionary convergence are very familiar: wings, for instance, have evolved independently at different times in different clades: insects, pterosaurs, birds, and bats all evolved wings at various times. In each case, the wings have rather different structures owing to the different body plans of the organisms. In pterosaurs, a single digit of the hand – a kind of elongated pinky finger – extends to support the wing. In bats, the wing membrane is supported by several elongated fingers. Insect wings are modifications of the animal's ecoskeleton. Natural selection solved the problem of flight in several different ways in these different groups. It's tempting to think that any land animals, with any body plans, would eventually have hit upon a way of taking to the skies given only enough evolutionary time. Insects did it. Reptiles did it. Birds did it. Mammals did it. The outcome looks inevitable, as if natural selection will always – somehow – find a way, no matter what starting materials it has to work with.

In his 2003 book, *Life's solution*, paleontologist Simon Conway Morris includes an impressive catalogue of examples of evolutionary convergence. Some of the examples are really mind-boggling, such as frogs and snakes living on different continents that secrete poisons having the same chemical structure. Whitehead (2008) describes another fascinating instance of convergence between elephants and sperm whales: although the two are only distantly related, they live in social groups that have remarkably similar structures. Both species are matriarchal. Young males leave their mothers' social groups and wander on their own, eventually growing to a larger size than the females. Later on, the male elephants and sperm whales return to move among the social groups they had earlier left behind, in order to find (or be found by) a mate. How strange that such different organisms should have such similar social lives.

Conway Morris (2003) argues that convergence is the hallmark of evolution. Not only that, but given enough time, evolution by natural selection will eventually produce creatures that are intelligent, communicative, self-aware, and have complex social lives. In other words, evolution will eventually produce creatures rather like us.

In the early 1980s, Canadian paleontologist Dale Russell wondered what would have happened if the non-avian dinosaurs had not gone extinct 65 million years ago. What if they had been granted all the time in the world to evolve? Russell had been studying the fossil remains of one dinosaur in particular – a small therapod called *Troödon* – and he was impressed by some of its features. To begin with, *Troödon* had a relatively big brain. Its encephalization quotient (EQ), which is the ratio of an animal's brain size to the average brain size of animals having a certain body size, was quite high. What's more, Russell noticed that at the very end of the Cretaceous, there was a trend toward higher EQ, but the trend was cut off abruptly by the K-T mass extinction event. Like humans, *Troödon* had good binocular vision, and it also seems to have had opposable fingers on its three-fingered hands. What might have happened had these small therapods been granted a reprieve and allowed to continue to evolve? Russell speculated that they would have evolved into intelligent "dinosauroids," and he even commissioned a sculptor to fashion an image of one, based on Russell's projection of evolutionary trends beyond the K-T extinction event. The view of evolution that is implicit in Russell's speculation is that convergence is to be expected. Given enough time, the evolution of intelligence is inevitable. Although it's tempting to pooh-pooh these speculations about human-like dinosauroids, the idea of convergence itself is not only not crazy, but is widely accepted in evolutionary biology. We already know of some perfectly good examples of convergence in different clades. Clayton and Emery (2008) observe that crows and apes have evolved similar kinds of intelligence and even tool use, even though their brains are structured in rather different ways.

In a critique of Conway Morris's argument, Sterelny (2005) worries that whether two traits qualify as convergent may depend, in part, on how we choose to describe them. One of Conway Morris's examples of convergence is agriculture. We humans are not the only creatures that practice agriculture: some leaf-cutter ants also maintain fungus gardens (Conway Morris 2003, pp. 198–200). Are the ants really practicing agriculture? Sterelny argues that it all depends on how broad or how narrow a view you take of agriculture. If you define "agriculture" more narrowly, then what the ants are doing and what we humans are doing will not count as the same thing, and therefore will not count as an instance of convergence. It might be possible for proponents of convergence to reframe their position, while taking Sterelny's criticism into account. Convergence, one might say, occurs when similar traits evolve in different lineages, from different evolutionary starting points. The greater the similarity, the greater the convergence.

Conway Morris's convergentism relates to a number of other issues in the philosophy of biology. First, in keeping with the adaptationist tradition, those who emphasize convergence may do so as a way of underscoring the power of natural selection. The standard explanation of evolutionary convergence, after all, is that different lineages have been subject to similar selective pressures over long periods of time. Second, Conway Morris's convergentism bears on the study of trends. Convergentists are generally favorable toward the idea that evolutionary history is full of trends that are driven by natural selection – think again of the increasing EQ in small therapod dinosaurs. At first glance, it might also seem that convergentists must believe in evolutionary progress. It's better, though, to keep these two ideas distinct. The claim that certain evolutionary outcomes are inevitable, and that evolutionary processes are insensitive to variations in initial conditions, is a descriptive claim about evolution. Believers in evolutionary progress must add the normative claim that those inevitable outcomes are good or desirable.

Replaying the tape of evolution

Conway Morris's 2003 book, *Life's Solution*, was a direct response to Stephen Jay Gould's *Wonderful Life* (1989). In that earlier book, Gould argued that history is highly contingent, and he used the following thought experiment to make the point:

> I call this experiment "replaying life's tape." You press the rewind button and, making sure you thoroughly erase everything that actually happened, go back to any time and place in the past – say, to the seas of the Burgess Shale. Then let the tape run again and see if the repetition looks at all like the original. (1989, p. 48)

If the outcome is different when you replay the tape, then the historical processes in question are contingent. Gould's thought experiment draws attention to our inability to perform the kinds of experiments with large-scale evolutionary processes that we might wish. But it is also supposed to help clarify the substantive claim that evolution, and perhaps history more generally, is contingent and unrepeatable.

Gould and Conway Morris also disagree about the place of human beings in the grand sweep of evolutionary history. Gould argues strenuously that we humans are the beneficiaries of a long string of historical accidents. If even the slightest thing had happened differently in the past, we never would have evolved at all:

> We came *this close* (put your thumb about a millimeter away from your index finger), thousands and thousands of times, to erasure by the veering of history down another sensible channel. Replay the tape a million times from a Burgess beginning, and I doubt that anything like *Homo sapiens* would ever evolve again. (1989, p. 289, emphasis in the original)

This view is diametrically opposed to that of Conway Morris, who thinks that the evolution of creatures rather like us was inevitable and fore-ordained from the very beginning.

Gould's claim that evolutionary processes are highly contingent has had a major influence on philosophers of biology. In the philosophy of biology, John Beatty (1995; 1997) has argued that the contingency of evolutionary history means that there are no strict empirical laws in biology. Beatty reasons that if there were any distinctively biological laws, they would have to be general-izations about the outcomes of highly contingent evolutionary processes, and hence would lack the kind of necessity that we expect genuine laws of nature to have. Traditionally, philosophers of science have held that a law of nature would have to be a true universal generalization. But philosophers have also fretted about examples of true universal generalizations that don't seem like *bona fide* laws. The logical empiricist philosopher C.G. Hempel (1966) imagined an alternative universe in which all the gold that exists is concentrated in one place, in the form of a giant sphere. In that universe, it would be true that if anything is an atom of gold, then it is part of a giant sphere. Intuitively, though, that does not seem like that would be a genuine law of nature. The generalization is true, but it seems to be true only by accident. Things might have been different. There is no reason why all the gold in the universe has to come in the form of a giant sphere. It seems like we want something more in a law of nature – namely, we want the law to tell us how things have to be. Thus, many philosophers distinguish between *genuine laws* and *accidentally true generalizations*.

Beatty argues along the following lines: if evolutionary history is con-tingent, as Gould claims, then all the living things that exist today might have been different. Now if there were any distinctively biological laws, those laws would have to be generalizations about living organisms. But if living things could have been completely different, those generalizations will only be true in virtue of historical accidents – just as we human beings only exist in virtue of historical accidents. Beatty's line of argument has given rise to a lively debate within philosophy of biology concerning biological laws (see e.g., Brandon 1997; Sober 1997; Mitchell 2003; Elgin 2006). One of the things at stake in this debate is whether biology is the same sort of science as physics and chemistry, which do have their own distinctive laws. Beatty thinks that the answer is no, and that the contingency of evolutionary history means that biology has no distinctive laws of its own. Beatty's work shows that there are important philosophical issues riding on the outcome of this debate.

At this point it will help to distinguish two different questions about evolutionary contingency. First, there is a *conceptual question*: What are some of the things that "evolutionary contingency" might mean? What might it mean to say that macroevolutionary processes are contingent? Second, there is an *empirical question*: How much contingency is there in evolutionary processes? These two questions are obviously related. How much contingency there is in evolution will depend on what we mean by "contingency" in the first place. I will take these questions in order: in the remainder of this chapter, I take up the conceptual question, and in the next chapter, I'll turn to the argument that Gould made in *Wonderful Life* based on his study of the fossils of the Burgess Shale.

Theological undertones

One other fascinating issue that Gould and Conway Morris disagree about is the relation between science and religion. I mention this issue here because I think their deeper disagreement about how to think about the relationship between science and religion informs their respective interpretations of the fossil record. Their science is being driven by deeper philosophical concerns.

Ian Barbour (1997) provides a helpful framework for thinking about how science and religion might be related. One view is the *conflict view*, according to which some of the findings of natural science conflict with some of the teachings of this or that religious tradition. The conflict might also be methodological. Some conflict theorists have held that a scientific approach to forming one's beliefs (whatever that might mean) is not compatible with certain kinds of religious commitment. A second view is the *independence view*.

According to this picture, science and religion occupy separate domains and engage with different kinds of questions about the world and our place in it. As long as everyone respects the proper boundaries, there could not be any genuine conflict between them. A third approach is the *dialogue view*, according to which science and religion might have something to teach one another. Science might actually reinforce some religious teachings, and in other cases it might serve as a corrective to religious excesses and super-stitions. By the same token, religious tradition can sometimes inspire new scientific ideas, and can serve as a check against certain kinds of excesses in science.

Gould was a leading spokesperson for the independence view. Indeed, he popularized the view that science and religion represent *non-overlapping magisteria (NOMA)*. As long as scientists restrict themselves to investigating how the world works, and as long as representatives of religious tradition stick with questions about ultimate value and meaning in life, no conflict is possible (Gould 1997b). Thus, for Gould, the question whether evolution is contingent should be taken as merely a scientific question, without any deeper religious ramifications. Indeed, on Gould's view, no scientific questions have any deeper religious consequences. Conway Morris does not see things the same way. He tends to think that science and religion have something to say to one another. Although he stops just short of claiming that convergentism supports the idea that the whole evolutionary process was created and sustained by God, he does seem to hint that the convergentist view of evolution coheres better with traditional theism than Gould's contingentist view does (Conway Morris, 2003, Chapter 11).

Conway Morris does have a point. From a certain perspective, it can look like Gould's own views about evolutionary contingency violate his view that science should be religiously neutral. A theist may well hold that God wanted to create a universe in which humans evolve via natural processes. Someone who goes this route must give up on a literal reading of the Bible but can take on board virtually everything that modern biology tells us about evolution. The crucial word there is "virtually." Could someone who seeks to combine evolution with theism in this way get behind Gould's claim that humans might never have evolved, and that we owe our existence to a long string of evolutionary accidents? At the very least, there is a tension between saying (a) that God wanted to create a world in which humans evolved by natural processes, and (b) that if we could rewind the tape of life and play it back again, humans probably wouldn't evolve at all. On the other hand, Conway Morris's convergentism fits much more comfortably with this attempt to combine belief in God with evolution. It is easy for someone who thinks that the evolution of human beings (or of creatures rather like us) was inevitable to see evolution as part of God's plan for the universe. The theistic religious traditions tend to be anthropocentric. They see the whole cosmic drama as being, in some sense, about human beings. Gould's insistence that evolution is contingent poses a threat to that anthropocentrism, and for that reason it may fail Gould's own test of theological neutrality.

The fact that Conway Morris has theological motivations for challenging Gould on the issue of contingency suggests that the dispute between them is as philosophical as it is scientific. Notice, though, that even if there is a tension between Gould's defense of contingency and his commitment to NOMA, the problem could lie with the latter, rather than the former.

Gould's two thought experiments

What does "contingency" mean? According to traditional philosophical usage, a contingent being is one that might not have existed, while a contingent truth is a proposition that might have been false. Philosophical tradition also recognizes different strengths of possibility, necessity, and contingency: logical possibility (or perhaps metaphysical possibility), physical possibility, biological possibility, technological possibility, and so on. This traditional way of thinking about contingency is non-historical. Gould's work, however, challenges us to think about whether there might be a distinctively historical kind of contingency. The proposals surveyed here are all attempts to spell out what "historical contingency" might mean.

A word of caution before we proceed: there is a process/product ambiguity that shows up in discussions of historical contingency. Sometimes, when scientists and philosophers raise the issue of contingency, they mainly want to know whether certain evolutionary outcomes might have been different. When the focus is on outcomes, talk of historical contingency comes much closer to traditional philosophical usage. When Gould says that our species (a product of evolution) is contingent, he might just be saying that we might not have existed. Presumably, though, the reason why we might not have existed is that evolutionary processes are, in some sense, contingent. If we really want to see what Gould was up to, we need to examine what it might mean to say that evolution (the process) is contingent.

When you replay the tape in Gould's thought experiment, one of two things can happen: either history will unfold in exactly the same way that it did originally, or else it will unfold in a new and different way. I assume that when you rewind the tape to a certain point in the past, you do not change the earlier conditions. In at least some versions of his thought experiment, Gould insisted that different outcomes can unfold even from the same starting point:

> We can explain an event after it occurs, but contingency precludes its repetition, even from an identical starting point. (1989, p. 278)

At other times, Gould seems to suggest that that contingency means that we get different outcomes from small variations in the initial conditions:

> Alter any early event, ever so slightly and without apparent importance at the time, and evolution cascades into a radically different channel. (1989, p. 51)

> Any replay, altered by an apparently insignificant jot or tittle at the outset, would have yielded an equally sensible and resolvable outcome of entirely different form. (1989, p. 289)

So do we, or don't we, alter the initial conditions before we replay the tape? Gould in effect offers two distinct thought experiments, which seems to indicate that he has two different senses of contingency in mind.

One might be forgiven for thinking that the first version of Gould's thought experiment – the one in which you replay the tape from exactly the same starting point – does little more than clarify the distinction between traditional philosophical determinism and indeterminism. After all, determinism is just the view that the history of the universe up to a given time t, together with the laws of nature, uniquely determines whatever happens after t. If you could replay life's tape and get a result that differs from the original one, that would show that determinism is false, as long as the earlier conditions and the laws of nature remain fixed. However, if "contingency" just refers to traditional philosophical indeterminism, then it is not entirely clear what paleontology might have to say about the issue. One challenge, then, is to try to figure out what else the first version of Gould's thought experiment might illustrate, if not philosophical indeterminism.

John Beatty on historical contingency

Beatty (2006) notices the two versions of Gould's thought experiment and argues that two different senses of "contingency" exist side by side in Gould's work: *contingency as unpredictability* and *contingency as causal dependence*. He goes on to make the intriguing suggestion that biologists sometimes fail to distinguish carefully between these two senses of "contingency."

Beatty insists that contingency as unpredictability "is not tantamount to any mysterious sort of indeterminism" (Beatty 2006, p. 345). Instead, he treats it as the view that "the occurrence of a particular prior state is *insufficient* to bring about a particular outcome" (2006, p. 339, emphasis in the original). The term "contingency as unpredictability" might be something of a misnomer. "Contingency as causal insufficiency" would be more felicitous. Whether an outcome is predictable or foreseeable will generally depend upon a whole host of contingent facts about the human beings doing the predicting, especially facts about their background knowledge, their inferential abilities, and so on. This does not seem to be what Beatty is talking about. Rather, he seems to want to interpret Gould as saying that in many cases, prior conditions do not guarantee the occurrence of any particular outcome. This sounds good at first, but what does Beatty mean by "prior conditions"? One possible reading is that the occurrence of a particular prior state *of the entire universe* is insufficient to bring about a given outcome. When Beatty's formulation is read in this way, contingency as causal insufficiency threatens to collapse again into indeterminism.

There is another more promising way to read Beatty's formulation of contingency as unpredictability. Consider the following example, which is borrowed from Cleland (2002): an electrical short occurs in a building in the middle of the night, causing the building to burn to the ground. But the short circuit, considered all by itself as a particular prior state, is causally insufficient for the burning of the building. A great many other conditions must obtain: there must be flammable materials nearby; the sprinkler system must fail; the fire crew must be delayed on the way to the scene; and so on. The electrical short is just one part of the total sufficient cause of the fire, just one of a package of jointly sufficient conditions. If contingency as causal insufficiency is understood in this way, it is compatible with causal determinism.

It is also possible to read Gould's thought experiment as an illustration of contingency as causal insufficiency in this weaker sense. Suppose we rewind the tape of history and hit "play" as soon as the electrical short occurs. The electrical short happens on every replay of the tape. However, if we vary other background conditions, we get different outcomes each time. Perhaps when Gould refers to playing the tape back multiple times "from an identical starting point," he just means that particular prior conditions must be held constant, and that those prior conditions are causally insufficient for producing a particular outcome.

One possible concern about this first sense of "contingency" is that it isn't very exciting. Contingency as causal insufficiency will show up wherever multiple causes contribute to bringing about some outcome. Nor is it very controversial to be told that history is contingent in this sense. Is there some way of making this notion of causal insufficiency more exciting? One way of reading Gould is to take him to be saying only that specific kinds of initial conditions are insufficient to bring about evolutionary outcomes. It is tempting to read him as saying that *natural selection* (or perhaps, selection plus ancestral state) typically does not guarantee evolutionary outcomes. This reading coheres with Gould's well-known critique of pan-selectionism (Gould and Lewontin 1979). Indeed, this is how Beatty reads him in the end:

> In denying that the same outcomes will result, Gould is not suggesting that the outcomes are inexplicable. Rather, he is denying that selection alone is sufficient to guarantee one particular outcome. (2006, pp. 341–342)

This interpretation of Gould also coheres nicely with the latter's discussion of the fauna of the Burgess Shale (Gould 1989). The creatures of the Burgess Shale had evolved a stunning variety of radically disparate body plans. At the end of the Cambrian period, most of those *Baupläne* disappeared forever. Only a few made it, including the humble *Pikaia*, a candidate for being the ancestor of all vertebrates. Perhaps Gould is claiming that selection alone did not guarantee this outcome. If we want to know why some lineages persisted beyond the Cambrian while others became extinct, we have to look at factors having nothing to do with natural selection.

Beatty does not carefully distinguish between contingency as causal insufficiency and contingency as the causal insufficiency of selection. The latter is clearly a special case of the former, so it is understandable that he would run them together. But the difference between the two could be important. Is contingency a feature of all historical processes, or is it only a feature of evolutionary processes? Focusing more narrowly on the causal insufficiency of natural selection does make contingency more exciting from the perspective of evolutionary theory, but there is a trade-off here. Gould's references to Tolstoy and the American Civil War in *Wonderful Life* make it clear that he thinks contingency is the hallmark of historical processes more generally, and not just a feature of evolutionary processes.

According to Beatty, there is another distinct sense of "contingency" that speakers typically have in mind when they say that one thing is contingent upon the occurrence of another. A later event B is contingent upon an earlier event A just in case A is causally necessary for the occurrence of B. This sense of contingency as *causal dependence* is clearly compatible with causal determinism. This is an all-or-nothing sense of "contingency"; a later event either depends on the occurrence of an earlier event, or it doesn't.

Beatty says explicitly that this notion of contingency as causal dependence "is intended in a way that is not trivial" (2006, p. 346). It may not seem very exciting or controversial to be told that later conditions in the history of life causally depend upon earlier conditions. However, Beatty's notion of causal dependence is closely related to the more interesting notion of sensitivity to initial conditions. Sensitivity to initial conditions is more interesting because it comes in degrees. At the one end of the spectrum it is possible to imagine cases where big changes in the initial conditions make no difference at all to the downstream outcomes. At the other end of the spectrum there are cases in which very tiny changes in the initial conditions make big differences to the outcomes. Philosopher Yemima Ben-Menahem (1997) also contrasts historical contingency with historical necessity or inevitability. The difference is one of degree: the greater the sensitivity to variations in initial conditions, the greater the contingency. When Conway Morris argues that evolution is highly convergent, what he means is that evolutionary outcomes are highly insensitive to initial conditions. Think again of the wings of birds, bats, pterosaurs, and insects: evolution produces similar outcomes from wildly different starting points.

Some readers come away from Gould's work with the thought that when he writes of contingency, he must be thinking about mass extinction events whose causes originate outside the system of evolving life on Earth. The classic example of this is the Cretaceous-Tertiary extinction event of 65 million years ago, which scientists today confidently attribute to an asteroid impact. Sensitivity to external disturbance is not quite the same thing as sensitivity to changes in initial conditions. In order to see why, it may help to reconsider the earlier example of the ball rolling down the side of a crater. Suppose that, just after the ball is released, a small earthquake occurs. This external disturbance might not change the outcome of the process; the ball will still come to rest at the bottom. Other kinds of disturbances might change the outcome. Systems can be more or less sensitive to these kinds of external disturbances.

If one wanted to, one could lump possible sources of disturbance together in the description of the initial conditions of a system. We may decide to draw the boundaries of the system however we want, and if we draw them broadly enough, sensitivity to external disturbances will show up as a kind of sensitivity to initial conditions. In practice, however, it is often useful to distinguish between these two ways in which the outcomes of historical processes can be sensitive to other factors. To the extent that we find this distinction useful, we might also find it useful to distinguish between two senses of contingency.

It might also be possible to generate an even richer sense of contingency by combining the notions of causal insufficiency and causal dependence. In fact, Beatty himself gestures in this direction:

> A historically contingent sequence of events is one in which the prior states are necessary or strongly necessary (causal-dependence version), but insufficient (unpredictability version) to bring about the outcome. (2006, p. 340)

We can call this kind of sequence a *Gallie-contingent sequence*, after W.B. Gallie, who discussed the idea in the late 1950s and 1960s. Gallie (1959; 1964) was primarily interested in whether there is a distinctively historical mode of explanation, which it was fashionable at the time to call "genetic explanation." Indeed, Gallie suggested that historians typically explain events by fitting them into what I am here calling Gallie-contingent sequences.

The above definition of a Gallie-contingent sequence might seem too weak to hold much interest. These kinds of sequences show up nearly everywhere we look. Consider, for example, the sequence of events involved in preparing a pasta dish. Filling the pot with water is necessary, but not sufficient, for boiling the water on the stove. My filling the pot with water is just part of the total sufficient cause of the water's boiling. Boiling the water on the stove is necessary, but not sufficient, for cooking the noodles. Cooking the noodles is necessary, but not sufficient, for producing fettucini alfredo. This is a simple Gallie-contingent sequence; at each step, there is both causal necessity and causal insufficiency. Thus, the claim that historical processes typically involve Gallie-contingent sequences does not seem too controversial. Perhaps that is why Gallie originally argued that offering an historical (genetic) explanation of some event requires more than merely fitting it into a Gallie-contingent sequence. The sequence must also have a trajectory; it must exhibit some sort of directional historical trend.

Nothing prevents us from trying to enrich the notion of a Gallie-contingent sequence by building in further requirements. One special case of a Gallie-contingent sequence would be an evolutionary process having the following features: (i) natural selection working on a population at an earlier time interval is not sufficient to determine the state of the population at the next time interval; (ii) the outcome of the process is highly sensitive to changes in the upstream historical conditions; and (iii) the process exhibits some sort of trajectory or directional trend. This formulation combines the notion of sensitivity to initial conditions with the notion of the causal insufficiency of selection. We might accordingly call a process that fits this description a *strongly* Gallie-contingent sequence.

Table 8.1 summarizes the different senses of "contingency" introduced so far. In the next section, I introduce a somewhat different proposal for thinking about what Gould might have meant by his thought experiment of replaying the tape, a proposal which integrates some of the ideas discussed up to now.

Contingency and the MBL model

Recall that during the early 1970s, Gould was involved with the MBL group, which devised the first computer simulation of large-scale evolutionary processes. The MBL model represented large-scale evolutionary processes as fundamentally stochastic. When Gould argues in *Wonderful Life* that evolution is highly contingent, it makes sense to read him as saying that macroevolution is a stochastic process – roughly the sort of process that is represented by the MBL model.

Although the term "stochastic" implies that a process is in some sense the result of chance, the outcomes of stochastic processes can be highly predictable. To take the simplest possible example, imagine a process in which someone tosses a fair coin 100 times. That process involves chance at every step, and yet it is easy to predict that the ratio of heads to tails will approach 50:50. The overall distribution of heads to tails is highly predictable, even if

Table 8.1 *Varieties of evolutionary contingency.*

		Pros and Cons
Contingency as unpredictability (Beatty 2006)	Causal insufficiency	Either collapses into philosophical indeterminism, or else is too weak because contributing causes, taken alone, are always insufficient to bring about effects.
	Causal insufficiency of natural selection	Applies only to evolutionary history, but has more theoretical interest. Related to the adaptationism debate. This is one of the things that Gould and Conway Morris disagree about.
Contingency as causal dependence (Beatty 2006)	All-or-nothing causal dependence	Boring. Causal dependence occurs everywhere in nature.
Contingency as sensitivity to external disturbance	Sensitivity to initial conditions (cf. Ben-Menahem 1997; Sterelny 2005)	More interesting, because sensitivity comes in degrees. Applies to all historical processes, not just evolution. This is something that Gould and Conway Morris disagree about. Relates to work on mass extinctions
Gallie-contingent sequences (Beatty 2006; Gallie 1959, 1964)	Weak sequences characterized by causal insufficiency and all-or-nothing causal dependence	Relatively boring. Such sequences occur everywhere in nature. Could serve as the basis for an account of historical explanation.
	Strong Gallie-contingent sequences in which selection is insufficient for later outcomes, and later outcomes are highly sensitive to earlier conditions.	The richest conception of contingency so far.

the result of a particular coin toss is not. This is a further reason why Beatty's talk of contingency as unpredictability has the potential to mislead.

The MBL model treats evolution as a stochastic process rather like a series of coin tosses. The computer simulation has the following 10 features, some of which pick up on earlier themes:

1. *Randomness.* What happens to each lineage from one turn to the next in the model, at least once the model reaches equilibrium, is random in the same sense in which the result of a coin toss is random. The different outcomes are equiprobable.

2. *Causal insufficiency.* The system exhibits contingency in Beatty's sense of causal insufficiency. The state of the system at turn 2 does not uniquely determine what state the system will be in at turn 3.

3. *Causal insufficiency of selection.* The MBL group explicitly set out to model phylogenetic branching processes in a way that leaves natural selection out of the picture. Natural selection's working on a lineage at one time interval doesn't determine what happens to that lineage in the next turn.

4. *Turn-by-turn unpredictability.* At turn 2, one can't predict what will happen at any subsequent turn.

5. *Predictability of patterns.* One can make predictions about certain patterns that will emerge over and over again with each run of the model. For example, one can predict that the diversity (the number of lineages) will increase until it reaches equilibrium.

6. *Sensitivity to changes in earlier states.* The states of the simulation at later turns depend on the states at earlier turns. Downstream conditions are highly sensitive to changes in upstream conditions.

7. *Sensitivity to changes in starting parameters.* Changing the parameters of the model – for example, the probabilities of the various outcomes or the number of lineages existing at turn 1 – will make a difference to the model outputs.

8. *Contingency of outcomes (in the traditional philosophical sense).* At the end of a model run, one can say that the evolution of a particular species is contingent, meaning that it might not have evolved.

9. *Rewindability.* A computer simulation such as this one is a lot like a tape that can be rewound and played back. Moreover, if you rewind the tape of the model to turn 2 and play it back (even without changing the initial conditions), you will very probably get a completely different outcome.

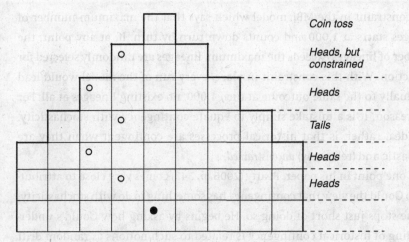

Figure 8.1 A constrained stochastic process.

10. *Gallie-contingency*. Stochastic processes, such as the branching process in the MBL model, are examples of Gallie-contingent sequences. But not all Gallie-contingent sequences involve stochastic processes.

Fascinatingly, the MBL model exhibits every interesting kind of historical contingency but one. The only variety of contingency that's not built into the model is sensitivity to external disturbance. This limitation may go with the territory of modeling. Recall the earlier point that the only difference between sensitivity to initial conditions and sensitivity to disturbance has to do with how we draw the boundaries around the system. If one were to write a computer simulation that builds in some sort of disturbance to the evolving system – say, if the programmers stipulate that 70% of the lineages, chosen at random, become extinct at turn 392 – it's not clear what it would mean to say that the model system is being disturbed from the outside.

In principle, it is possible for a stochastic process to have an inevitable outcome. This could happen, for example, if the state space is structured in a certain way. Consider the simple coin tossing game shown in Figure 8.1. Let's stipulate that if the coin comes up heads, the marker moves north one space and west one space. If tails, the marker moves north one space and east one space. If the marker hits the northwest edge of the grid, it remains in place until the coin comes up tails, and similarly for the northeastern boundary. Clearly, every time we play this game, the marker will end up on exactly the same square. The outcome is inevitable, even though the process of getting there is a stochastic one. Similarly, it would be possible to build in a constraint in the MBL model which says that the maximum number of lineages starts at 1,000 and counts down turn by turn. If, at any point, the number of lineages exceeds the maximum, lineages are randomly selected for extinction. With this constraint in place, every run of the model would lead eventually to the same outcome at turn 1,000: no existing lineages at all. For this reason, it is a mistake simply to equate contingency with stochasticity. The idea, rather, is that historical processes are contingent when they are stochastic and (relatively) *unconstrained*.

At one point in his paper, Beatty (2006, p. 345) comes very close to attributing to Gould the view that contingency has something to do with stochasticity, but he stops just short of doing so. He begins by asking how Gould's understanding of historical contingency is related to such notions as random drift and random mutation. He identifies a possible tension in Gould's view: on the one hand, he cites a paper (Gould and Woodruff 1990) in which Gould wants to distinguish between historical contingency and stochastic effects such as drift. At the same time, Beatty (who at this point in his paper is thinking mainly about contingency as the causal insufficiency of selection) suggests that Gould is also committed to thinking of drift and mutation as sources of historical contingency. Beatty then says:

> I cannot go on without noting that this distinction – between stochastic effects like random drift due to sampling error, on the one hand, and historical contingency, on the other – does not square very well with Gould's own account of the persistence of some forms rather than others following the Cambrian explosion. According to Gould, that outcome was highly contingent and could well have turned out otherwise because it had so little to do with differences in adaptive fit and was more like a lottery. Which sounds a lot like sampling error at the level of lineages. (2006, p. 345)

On this reading, evolutionary contingency is the macro-level analogue of drift; it is just random species sorting. Of course, the MBL model treats species sorting as random. Beatty is certainly on the right track here, although he misses the connection with the work of the MBL group in the 1970s. Beatty also does not quite recognize – though he is very close – that there is a distinct sense of "contingency" here. This notion of contingency as (unconstrained) stochasticity is distinct from the other two interesting notions which Beatty discusses and takes to be central – the causal insufficiency of selection and sensitivity to initial conditions – because it is *stronger* than either of those two notions taken alone and incorporates both of them.

In an important popular essay ("The wheel of fortune and the wedges of progress") that further develops some of the themes of *Wonderful Life*, Gould distinguishes between two different ways of thinking about the elimination of lineages during mass extinction events (1993a). The first is the *random model*, according to which "species live or die by the roll of the dice" (p. 307). The second is the *different rules model*. According to this second model:

> Species live or die for definite specifiable reasons. But the causes of success are quirky and fortuitous with respect to initial reasons for evolving the features that secured survival. (p. 307)

The random model is the one that treats large-scale evolution as a genuinely stochastic process. It is the more radical of the two, and Gould says quite explicitly that he "sees a role" for the random model in understanding certain mass extinction events.

Does this notion of contingency as unconstrained stochasticity threaten to collapse once again into indeterminism? Earlier on I agreed with Beatty that we should resist thinking of historical contingency as synonymous with causal indeterminism as traditionally understood by philosophers. Indeed, Gould himself insists that "contingency is not the titration of determinism by randomness" (1989, p. 51). So how does this notion of contingency as unconstrained stochasticity differ from causal indeterminism?

To begin with, notice that if we think of macroevolutionary contingency as unconstrained stochasticity, then it will have the same status as random genetic drift at the level of microevolution. No one thinks that belief in random genetic drift comes with any commitments one way or the other with respect to the traditional philosophical question whether determinism is true. Belief in macroevolutionary contingency (as unconstrained stochasticity) is no more or less problematic, from a metaphysical perspective, than belief in random drift at the micro-level. More generally, it is uncontroversial that stochastic processes can take place in a LaPlacean deterministic world (cf. Millstein 2000). For example, even if LaPlacean determinism is true, we can still conduct a trial of 100 tosses of a fair coin. It is an interesting question what it might mean to say, in a LaPlacean deterministic world, that the probability of obtaining heads is .5. But at least two of the traditionally popular interpretations of probability – the epistemic and the frequentist interpretations – are perfectly compatible with LaPlacean determinism. Whether the propensity interpretation is also compatible with determinism is more complicated (see e.g., Mellor 2005, Chapter 4). Thus, the claim that macroevolution is a stochastic process comes out neutral with respect to the traditional philosophical problem of determinism *vs.* indeterminism, depending on how one interprets probability theory.

Gould repeatedly distinguished historical contingency from randomness (see e.g., 1989, p. 283). At first glance, this would seem to weigh strongly against the view that by "contingency" he really meant macro-level stochasticity. But what exactly did Gould mean by "randomness"? In another popular essay (1993b, pp. 396–397), Gould also carefully distinguished between different senses of "randomness" – a scientific and a vernacular sense. I would suggest that when he wrote of the "titration of determinism by randomness" (1989, p. 51), he was using "randomness" in what he thought of as the vernacular sense. In the vernacular, the word "random" connotes disorder, unpredictability, and the absence of pattern, and perhaps even the absence of causation. But Gould insisted (1993b, p. 396) that processes which are random in the scientific sense (i.e., stochastic process) can generate predictable order and pattern. That's why we should "pray for randomness" if we want to understand history.

This notion of contingency as unconstrained stochasticity is closely connected to some of the questions about the relationship between micro- and macroevolution that we examined in Chapters 4 and 5. In those chapters we focused mainly on the notion of species selection. The early proponents of species selection had the ambitious goal of developing a theory of macroevolutionary change that would parallel the modern synthetic theory of microevolution. The rough parallel was meant to look something like the following:

Natural selection	Species selection
Mutation	New species appear randomly with respect to trends in the morphospace
Random genetic drift	Random/unbiased species sorting

On this reading, the debate about contingency is – like so much else in modern paleontology – really a debate about the hierarchical expansion of evolutionary theory. The claim that evolution is contingent boils down to the claim that species sorting is unbiased. Seeing things this way also helps make clear why Gould thought he could coherently argue that macroevolution is contingent without denying traditional causal determinism. Random or unbiased species sorting in no way contravenes traditional determinism.

I've now distinguished no fewer that eight senses of "Evolutionary contingency." You may be wondering which one of these is the *right* one? Unfortunately there is no simple answer to that question. Like Beatty, I strongly suspect that Gould used the term "contingency" in different ways at different times, even within the same book. Conway Morris, in opposition to Gould, will surely deny that evolution exhibits much contingency in several of these senses of contingency. For my part, I would argue that several of these different senses of contingency are especially interesting, including sensitivity to initial conditions, sensitivity to disturbance, the causal insufficiency of selection, and the notion of random species sorting. Notice, though, that these distinctions seriously complicate the empirical investigation of contingency.

An experiment in evolution

In this chapter, I've focused mainly on the conceptual question: What is evolutionary contingency? But that leaves the empirical question: How contingent is evolutionary history? Of course we can't really say what the empirical question means until we clarify which sense of contingency we are talking about. How might scientists go about answering the empirical question, once that question is suitably clarified? We have seen that Conway Morris (2003) argues against contingency by offering a catalogue of stunning evolutionary convergences. Although the catalogue is impressive, we have seen that there is a problem concerning how to count convergences (Sterelny 2005). In addition, it's not clear what conclusions one should draw about evolution from a list of examples. The debate about contingency is another example of what Beatty calls a relative significance debate. The friends of contingency would never deny that convergence is a significant evolutionary phenomenon; they just think that contingency is more significant.

Richard Lenski and his lab team at Michigan State University have tried to tackle the empirical issue by designing a set of experiments that really do "replay the tape" of evolution (Travisano et al. 1995). In one experiment they created twelve genetically identical strains of *E. coli* bacteria. These populations then evolved in separate petri dishes for 2,000 generations in a glucose-limited environment. The scientists also stored some of the original bacteria in a freezer. After 2,000 generations, they measured the fitness of the evolved bacteria from each of the twelve populations by letting them compete against their thawed-out ancestors. It turns out that the populations were about equally fit with respect to the glucose-limited environment in which they had been evolving. Fitness had increased in each of the twelve populations, and there were no major differences among the populations – clear evidence of natural selection at work. In essence, all twelve populations became better adapted to life in an environment where glucose is the only food. But then the researchers played a trick on the bacteria and switched sugars. They now let the evolved bacteria compete against their ancestors in a maltose-limited environment – that is, in a new environment different from the one in which they had spent the last 2,000 generations. This time, there was considerable variation. Some populations showed a higher fitness than others in the maltose-limited environment. The scientists attributed these differences to random drift and mutational ordering – in other words, to microevolutionary change.

Next, Lenski's team changed the experimental set-up. They took one individual from each of the twelve populations and cloned each one three times. This enabled them to create thirty-six new populations representing twelve different genotypes. They let these populations evolve for another 1,000 generations in a maltose-rich environment. They were, in effect, playing the tape of evolution from twelve slightly different starting points, because the twelve populations had different fitnesses relative to the maltose-limited environment. After 1,000 generations, they once again let the bacteria compete against their ancestors and measured their fitness relative to maltose-limited environments. Fitness once again increased in all the populations as they adapted to the new environment. In addition, the variation across populations was considerably reduced after 1,000 generations of evolution. The different populations were converging on – or, if you will, independently "discovering" – more efficient ways of metabolizing maltose. Although the experiment was inspired by Gould's work, this seems like a clear victory for convergence.

It's not clear, however, just how much we should make of this experiment. It certainly looks like an instance of evolutionary convergence, so we might well add it to Conway Morris's already impressive list. However, in a relative significance debate, one example will not tip the balance one way or the other. Furthermore, Lenski's group was studying a microevolutionary process. It's not clear what their observations of E. coli bacteria can tell us about larger-scale historical processes. The debate between Gould and Conway Morris is a paleontological one. They are concerned about the grand sweep of the history of life on earth, and especially with questions about the evolution of plants and multicellular animals. We need some further reason to think that what's true for *E. coli* at small scales will also be true for plants and animals at much larger scales.

Finally, one of the most interesting senses of contingency is the last one: contingency as stochasticity, or as random species sorting. Unlike the other senses of contingency, this one is distinctively macroevolutionary. Lenski's experiments do not really address the question whether evolution is contingent in this last sense. That is no fault of the experiments; for no microevolutionary experiment can tell us how much contingency there is in evolution, when the sense of contingency we care about is distinctively macroevolutionary.

9 Diversity, disparity, and the Burgess Shale

Does the fossil record have anything to say on the issue of contingency *vs.* convergence, beyond supplying examples of evolutionary convergence? In this chapter, I examine the main argument that Gould develops in his 1989 book, *Wonderful Life*. I also look briefly at recent reassessments of that argument (Brysse 2008; MacLaurin and Sterelny 2008; see also Baron 2009). In the end, I'm not optimistic that the fossil record will be able to settle what may well be the deepest questions about evolution. Gould's argument, as it happens, is rather difficult to make out. His discussion of the Burgess Shale organisms raises some important philosophical questions – for instance, about the concept of disparity and about methods of biological classification – but it is a challenge to figure out why Gould thought that the fossil record might contain evidence that evolution is contingent. Belief in the contingency of evolutionary history is something that scientists bring to the fossil record, rather than something they extract from it.

Gould on the Burgess Shale

The Burgess Shale, in the Canadian Rockies, is what scientists sometimes call a *Lagerstätte* – a layer of rock containing a high concentration and large variety of fossil remains. Discovered in 1909, the Burgess Shale dates from around 505 million years ago, and it offers scientists a vivid look at invertebrate marine life during the Cambrian period, before the evolution of vertebrates, and before life had moved from the oceans onto the dry land. The Burgess fossils were first described by Charles Doolittle Walcott of the Smithsonian Institution. Some of the fossils of the Burgess Shale are very strange, to say the least, and classifying them proved challenging. Walcott adopted the most straightforward approach, by trying to "shoehorn" – that's Gould's term – the Burgess creatures into existing animal phyla. Scientists typically think of each phylum as having

Figure 9.1 *Marrella splendens*. Reprinted with permission from the Smithsonian Institution – National Museum of Natural History.

a distinctive body plan, or *Bauplan*. For example, annelid (or segmented) worms represent one phylum, as do mollusks, and echinoderms (starfish and sea urchins). All vertebrates fall into the phylum Chordata. Insects, shrimp, crabs, lobsters, spiders, and scorpions are all arthropods. When Walcott approached the Burgess Shale creatures, he assumed going in that they must belong to one or the other of these existing phyla. Moreover, some of the Burgess creatures looked a bit like arthropods, but not quite like any living animals belonging to that group. In those cases, Walcott "shoehorned" the Burgess animals into one of the recognized classes of arthropods, such as the crustaceans (including shrimp, barnacles, and crawfish), or the trilobites (which are now extinct).

Just to give an idea of Walcott's procedure, Figure 9.1 shows one kind of fossil that he named *Marrella splendens*. Walcott just classified *Marrella* as a trilobite, a well-defined group of crustaceans, even though *Marrella* has a number of features that make it unlike any other trilobite. Just to give an example, *Marrella*, like other arthropods, has jointed limbs. However, its legs have six segments, whereas virtually all trilobites have legs with eight segments. *Marrella* also has weird spines on its head, which are not seen in any other trilobites. *Marrella* has two pairs of antennae on the front of its head, whereas trilobites have only one pair. Although *Marrella* looks like an arthropod, it didn't seem to fit neatly into any arthropod class, so Walcott just decided to call it a trilobite.

One of the most bizarre organisms of the Burgess Shale is *Opabinia* (see Figure 9.2). It has a segmented body with gills on top, a long nozzle-like feeding

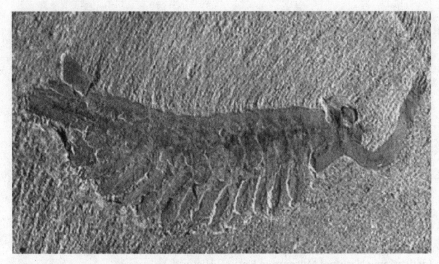

Figure 9.2 *Opabinia*. Reprinted with permission from the Smithsonian Institution – National Museum of Natural History.

appendage in front, and five eyes. Is it an arthropod? Walcott classified it as a crustacean, even though it lacks some of the distinctive features of arthropods, such as antennae and legs. Walcott apparently thought that the legs might have been squashed under the bodies in the fossils he was looking at. In the first phase of study of the Burgess Shale fauna from the 1910s to the 1930s, many other animals received a treatment similar to that of *Marrella* and *Opabinia:* they were assigned to well-recognized invertebrate taxa, even in cases where they didn't seem to fit very well.

In the 1960s and 1970s, a team of paleontologists led by Harry Whittington, of Cambridge University, took a second look at the Burgess animals. Ironically, one of the scientists involved in this effort was Simon Conway Morris, who would later go on to oppose Gould's views about evolution. Whittington had the benefit of improved techniques for studying fossils. For example, he and his team sliced up some of the abundant Burgess specimens and looked at cross sections. In some cases, they also had better specimens to study. Their work represents what Brysse (2008) calls the second phase of research on the Burgess organisms. Whittington, Conway Morris, and Derek Briggs emphasized the sheer strangeness of the Burgess creatures, and in many cases moved them out of the phyla and classes that Walcott had placed them in. Whittington, while allowing that *Marrella* was probably an arthropod, argued that it could not possibly have been a trilobite. He also showed that *Opabinia* was not only not a crustacean, but that it did not belong in the arthropod phylum at all. Cross-sections revealed that it lacked any of the usual appendages associated with arthropods. It had gills and lobes on its segmented body, but they were not even attached to the body in the same way as arthropod limbs. *Opabinia* seemed to have a unique body plan all its own – a little bit like an arthropod, and a little like a segmented worm, but quite distinct from both. Indeed, it seemed like a phylum unto itself. Writing in 1989, towards the end of phase 2 of the work on the Burgess Shale fauna, Gould concluded that there must have been many more animal phyla around during the Cambrian than there are today. Indeed, he argued that by most estimates, there are some 20–30 animal phyla around today. However, Whittington and his colleagues had described organisms from the Burgess Shale that were so weird, and had such unusual body plans, that we should conclude that there were 15–20 completely different animal phyla around during the Cambrian, all of which became extinct. And in addition to the currently recognized classes of arthropods – Uniramia, Chelicerata, Crustacea, and Trilobita – scientists would need to add more than twenty new classes to accommodate animals such as *Marrella*, which are obviously not trilobites, and obviously not crustaceans. Thus, one of the central claims of Gould's book involves classification: back during the Cambrian period, there were a lot of animal phyla, and a lot of arthropod classes, that subsequently became extinct.

This, it turns out, is just the beginning of the story, because by the time Gould published his book, some scientists had already begun to rethink the Burgess Shale yet again, this time in the light of the cladistic approach to biological taxonomy (Brysse 2008). Before we look at how Gould's argument has fared since the early 1990s, we first need to see how his point about the classification of creatures such as *Marrella* and *Opabinia* is related to the issue of contingency. And before we can do that, we need to consider the crucial distinction between diversity and disparity.

Diversity and disparity

Biologists typically think of diversity in terms of species richness, though in paleontology it is also common to measure diversity by counting genera and families. Changes in biological diversity over time have long interested paleontologists. Increasing diversity is one of the large-scale trends that paleontologists investigate. Gould (1989, p. 49) argued that there is an important distinction to be drawn between diversity and disparity, where disparity has to do with how different organisms are from one another, or "difference in body plans." Consider, for instance, beetles. Around 350,000 species of beetles have been described, making that group, the coleoptera, by far the most diverse group of animals. But beetles have low disparity, in Gould's sense. Although they exhibit a lot of variation in size, coloration, and in the shapes and proportions of their appendages, they all have the same basic body plan: a head, thorax, and abdomen, six legs attached to the thorax, hard, shelly forewings, and so on. All 350,000 species are variations on the same architectural theme. Gould argued that in the Burgess Shale, species diversity is relatively low, but that disparity was extremely high. Indeed, based on the reclassification of the Burgess organisms by Whittington and his team, Gould contended that there has never been any other time in the history of life on Earth when more animal phyla existed in the same marine ecosystem. Not only that, but the arthropods also hit their peak of disparity during the Cambrian. At no time since 505 million years ago have there been so many different kinds of arthropods swimming around, with so many different body architectures.

Why did Gould think that this distinction between diversity and disparity is so important? His emphasis on the importance of biological disparity has to be understood in the context of his rejection of the notion of evolutionary progress. Gould attacked the notion of progress over and over again, from a variety of different scientific and philosophical angles; this is just one of those angles. In the 1980s, paleontologists had focussed quite a lot on patterns of bio-diversity change in the fossil record. One interesting debate was between those who thought that biodiversity usually increases exponentially, with periodic interruptions by mass extinction events, and those who thought that biodiver-sity change follows more of an S-curve, increasing rapidly but then leveling off after a while once the number of species approaches equilibrium. Indeed, in one of the early triumphs of the new statistical paleobiology, Jack Sepkoski did a series of studies of the diversification of marine invertebrates and found that it followed an S-curve (Sepkoski 1978, 1979, 1984; for a helpful discussion of Sepkoski's work, see Ruse 1999). Although diversity was getting a lot of atten-tion, no one ever really questioned the assumption that biological diversity increases over evolutionary time. The disagreements all concerned the tempo and the causes of increase. And let's be honest, everybody knows that biodiver-sity is a good thing. In Chapter 7, I defended the view that when we make judg-ments about evolutionary progress, we inevitably bring our own standards to bear. By the 1980s, many people had become concerned about the rapid loss of biological diversity due to habitat destruction and other human activities. Environmentalists had begun to sound the alarm about the potentially bad consequences of rapid biodiversity loss. And some paleontologists, including Sepkoski, explicitly connected their work on the history of biodiversity with current concerns about human-caused extinction. In that cultural context, it may seem natural to think of biodiversity increase as a kind of evolutionary progress. Gould's rather ingenious move was to try to shift the focus away from diversity and onto disparity. He readily conceded that biological diversity has increased over time, but he argued that exactly the opposite is the case with respect to disparity. Disparity was dramatically reduced by extinctions at the end of the Cambrian, and it has never recovered. The biologically diverse world that we inhabit today remains, in a sense, impoverished. Some basic body plans – like that of the beetles – have seen a staggering increase in diversity, but the overall number of body plans is much reduced. At the very least, this challenges naïve assumptions about progress. Maybe disparity, like diversity, is something we should value. If so, evolutionary history is anything but progressive.

One potential worry about Gould's argument concerns the notion of disparity itself. What does that notion really amount to? In Chapter 7, I introduced the concept of a morphospace, a kind of space of biological possibility. Intuitively, we might think of disparity as having to do with the size of the region of the morphospace that living things occupy. The large number of distinct body plans that existed during the Cambrian meant that, at that time, living things were taking up a big swath of the space of possible organismic designs. Today, living things occupy a much smaller region of the morphospace. Indeed, those 350,000 species of beetles are all clustered together in one relatively small area. So one proposal is to define biological disparity in terms of the notion of a morphospace.

MacLaurin and Sterelny (2008) raise some doubts about this move. Although the concept of a morphospace is perfectly respectable, and indeed is widely used by biologists and paleontologists, MacLaurin and Sterelny argue that the notion of a morphospace only really makes sense when applied in a localized way. In order to see why, think about how a morphospace would have to be constructed. Every "space" has to have a certain number of dimensions. When creating a morphospace, you have to assign a dimension for each trait you are interested in. Back in Chapter 7, in the context of discussing evolutionary trends, we focused on models of clades evolving in a certain direction through a state space. But the state space, in that case, only involved one trait – body size! Other scientists work with state spaces having more dimensions than that. For example, Niklas (1997; 1998) uses a computer simulation to model the evolution of early land plants. In Niklas's model, plant lineages evolve by moving through a morphospace, but the morphospace is defined using only three parameters. Niklas is especially interested in the architecture of plants, so he idealizes away from most of the features that plants have in order to focus on the number of places where the stem bifurcates, the angle of bifurcation, and the angle of rotation. You could easily measure these three quantities for any living plant and place it at a certain point in the three-dimensional morphospace. There is a simple reason why Niklas focuses on only three traits: a four or higher dimensional morphospace becomes vastly more difficult to represent, and also more difficult to wrap one's mind around. MacLaurin and Sterelny object that in order to make his argument about the overall disparity in the Cambrian, Gould needs to help himself to the notion of a global morphospace, one that is defined with respect to a large number of traits. If you only picked three traits and then tried to argue that there was more disparity, with respect to those three traits, in the Cambrian than at any other time, which three traits would you choose? In describing the differences between *Opabinia* and arthropods, one needs to mention a large number of traits: eye number, presence or absence of antennae, number of body segments, presence or absence of walking legs, and so on and on. Three traits won't be enough to capture the disparity of the Burgess fauna. However, MacLaurin and Sterelny are skeptical that thinking about a global morphospace helps to clarify anything. Saying that the Burgess fauna occupied a large region of the global morphospace is no more illuminating than saying that they had a lot of disparity. The former concept does not help to make the latter more precise.

MacLaurin and Sterelny do not deny that one can, in principle, define a global morphospace in a highly abstract way. Dennett (1995), inspired by Jorge Luis Borges's short story about the Library of Babel, defines the "Library of Mendel" as a three-dimensional space containing all possible genomes. Such a space would be unimaginably large. For example, one ridiculously large shelf within the Library of Mendel would consist of my dog's genome, plus every possible genome that one could create by changing just one nucleotide somewhere on one of my dog's chromosomes. By analogy with Dennett's space of all possible genomes, one could simply say that the global morphospace is the space of all possible morphologies. The problem, though, is that talking about such a space does nothing to make the notion of disparity more precise.

MacLaurin and Sterelny also think that the problems with the notion of disparity are related to the problems with the so-called *phenetic* approach to taxonomy. For a while, phenetic taxonomy was a serious rival of the cladistic approach, which I will discuss shortly. The pheneticists sought to classify organisms on the basis of overall similarity. What makes it the case that you and I belong to the same species is that, overall, we are quite similar to one another. We are more similar to each other than either of us is, say, to a dog. The pheneticists also applied this approach to the grouping of species together into higher taxa. Why is it that humans and chimpanzees are both apes, while baboons are not? The simplest answer, the pheneticists thought, is that humans and chimps are more similar to each other, according to some overall measure of similarity, than either group is to baboons. One possible advantage of the phenetic approach is its theory-neutrality. It enables one to do biological systematics (that is, classification) without making any assumptions about the processes by which the different taxonomic groupings came about.

One fairly obvious problem with the phenetic approach concerns organisms whose lifecycles involve very different developmental stages. The adults of one butterfly species, intuitively, seem very different from the larvae of the same species. They might well seem more similar to the adults of the next species of butterfly than to the larvae of their own species. A more general problem – and this is the one that eventually sank phenetic taxonomy – is that no one could say how to measure overall similarity. The only way to do so is to come up with a list of different traits, and then to compare organisms with respect to those traits, but which traits should you use? And should all traits be given equal weight? Even worse, the very notion of a trait is somewhat subjective, because we can measure and describe anything that strikes our fancy. Think for a moment about how we talk about our own traits. We focus a lot on traits that seem relevant to human health: think about body mass index, blood pressure, and cholesterol levels. We could measure those sorts of things in other animals as well, if we cared to. Should those traits be included in pheneticists' measures of overall similarity? How do you decide which traits to include and which to ignore? Unfortunately, the notion of disparity also seems to presuppose some kind of measure of overall similarity. The term "disparity" in fact just seems like a synonym for "overall dissimilarity." The failure of pheneticists to provide a clear way of thinking about overall similarity should perhaps make us skeptical of the usefulness of Gould's notion of disparity.

It might be possible to defend Gould here by arguing that his critics are demanding too much. One strategy would be to point out that the concept of biodiversity is not really any clearer at the end of the day than the concept of disparity. The concept of biodiversity seems straightforward enough; all you need to do in order to measure biodiversity is to count species. However, counting species is vastly more complicated than it might seem. One problem is that there are several species concepts to choose from – i.e., several different accounts of what makes it the case that two organisms belong to the same species – and different scientists using different species concepts will arrive at different species counts. Earlier on I mentioned that Ernst Mayr's biological species concept is one of the most popular. It seems relatively straightforward to count species based in the criterion of reproductive isolation. But reproductive isolation also comes in degrees. Grizzly bears and polar bears don't usually hybridize in nature, but they can. Should they be considered two species, or one? The big problem with the notion of biodiversity, however, is the microbial world. Most of the diversity of life on Earth is microbial – everyone agrees about that much – but just how to group single-celled organisms into species is a matter of ongoing controversy. When we bring the microbial world into consideration, the notion of disparity does not seem any less precise, or any less muddled, than the notion of species diversity.

From the Burgess Shale to evolutionary contingency

Even if we grant Gould's point that disparity was radically reduced at the end of the Cambrian, it's not obvious how that fact counts in favor of the thesis that evolution is highly contingent. What is the connection?

Gould pointed out that one of the humbler organisms of the Burgess Shale – *Pikaia* – is at least a candidate for being an ancestor of all vertebrates. With that in mind, he asked: Why is it that some of the body plans represented in the Burgess Shale persisted, while others disappeared? What could possibly explain the difference? There was obviously some kind of sorting process that occurred. What, if anything, can we say about the nature of that process? *Pikaia* quite possibly gave rise to all subsequent vertebrates, but the strange body plan of *Opabinia* was lost to evolution forever. What accounts for the different fates of these two *Baupläne*? Gould also explicitly ties his thought experiment of "replaying the tape" to this discussion of Cambrian invertebrates. If we could rewind evolution back to the Cambrian, when all those disparate body plans still existed, and play it back, what would happen?

One live hypothesis is that the sorting of body plans at the end of the Cambrian was random. "Random" in this context does not mean "uncaused"; rather, it just means that there was nothing about *Pikaia* that gave it a higher probability of persistence than *Opabinia*. The sorting of body plans was unbiased. Gould puts the point quite vividly in the following passage:

> [I]f a radical decimation of a much greater range of initial possibilities determined the pattern of later life, including the chance of our own origin, then consider the alternatives. Suppose that ten of a hundred designs will survive and diversify. If the ten survivors are predictable by superiority of anatomy (interpretation 1), then they will win each time – and Burgess eliminations do not challenge our comforting view of life. But if the ten survivors are protégés of Lady Luck or fortunate beneficiaries of odd historical contingencies (interpretation 2), then each replay of the tape will yield a different set of survivors and a different history. (1989, p. 50)

Interpretation 2 is consistent with my suggestion in Chapter 8 that Gould is thinking of contingency mainly in terms of stochasticity, or random sorting at large scales. He is approaching the Burgess Shale with the MBL model in mind. If the sorting of body plans was unbiased, and if we could replay the tape (think: re-run the computer simulation), then it would be highly improbable that we would obtain the same outcome on two successive trials.

What's missing from Gould's argument in *Wonderful Life*, as far as I can tell, is any clear defense of interpretation 2. As we saw in earlier chapters, macro-level sorting can be either biased (interpretation 1) or unbiased (interpretation 2). There is a crucial, irredeemably speculative step in Gould's reasoning. He has to make something like the following inference:

> No one who studies the Burgess fauna can think of any reason why one body type have a higher probability of survival than another.
>
> Therefore, the sorting of body types was (probably) random.

This inference commits what logicians call the fallacy of ignorance. It amounts to saying: we don't know whether there were any biases in play; therefore, there weren't any. Gould comes very close to committing this fallacy, and perhaps does commit it, in his epilogue on *Pikaia*, the humble worm-like creature that might (or might not) have been the ancestor of us vertebrates:

> Wind the tape of life back to Burgess times, and let it play again. If *Pikaia* does not survive in the replay, we are wiped out of future history – all of us, from shark to robin to orangutan. And I don't think that any handicapper, given Burgess evidence as known today, would ever have granted very favorable odds for the persistence of *Pikaia*. (1989, p. 323)

That last line comes very close to saying that no one can tell, given only the fossil evidence, what *Pikaia*'s probability of persistence might have been. That may be true, but it doesn't logically imply that sorting was random.

Cladistics and the second reclassification of the Burgess Shale organisms

Increasingly, it looks like Gould's argument for evolutionary contingency may not withstand the test of time. Philosopher of paleontology Keynyn Brysse (2008) points out that research on the Burgess Shale continued after Gould published *Wonderful Life*. In this third phase of research on the Burgess Shale creatures, scientists used cladistic methods of taxonomy to rethink the relationship between the "weird wonders" of the Burgess Shale and existing groups.

Cladism represents a stripped down approach to biological taxonomy. According to this view, which was first proposed by Willi Hennig in the 1950s and has since gained ascendancy, every taxon – that is to say, every grouping of species – should be a clade, or what is known as a *monophyletic group*. A monophyletic group consists of a species plus all and only the other species that are descended from it. Monophyletic groups are to be distinguished from paraphyletic groups, which contain a species plus some (but not all) of its descendants, as well as polyphyletic groups, which contain a species plus its descendants and some other species besides. Cladism turns out to be quite a radical approach. When it first burst on the scene, in the 1950s and 1960s, biologists had of course already been in the business of classifying organisms for a very long time, and many well-known groupings were deeply entrenched. (See Hull 1988 for an account of the controversies surrounding the rise of cladism.) For example, we all think we know what reptiles

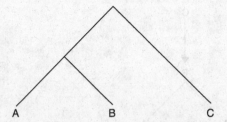

Figure 9.3 A cladogram. This diagram represents the historical hypothesis that species A shares a more recent common ancestor with B than with C.

are – they include snakes, lizards, and turtles as well as many extinct groups, such as dinosaurs, mosasaurs, and pterosaurs. However, it is now quite well established that birds evolved from small therapod dinosaurs. Even mammals evolved from the mammal-like reptiles of the Permian. The term "reptiles" therefore refers to a paraphyletic group consisting of a distant ancestor from the Paleozoic and only some of the groups that descended from that ancestor. A tough-minded cladist would argue that the concept of a reptile has no place in modern biological theory. Reptiles represent a merely artificial grouping of species. The view that living birds are dinosaurs is also associated with cladism. If dinosaurs are to constitute a monophyletic group, then the group must include everything descended from the earliest dinosaurs that evolved back in the Triassic, and that includes birds. In short, cladists hold that the purpose of biological taxonomy is to capture information about evolutionary relationships.

Cladism reflects a kind of theoretical trade-off. On the one hand, cladists have attained a high degree of methodological rigor, and they have an elegant view of what taxonomy is all about. However, they achieve those gains by narrowing their sights and leaving aside many questions about evolution that one might find interesting. Cladists focus primarily on the following sort of question: Given three species, A, B, and C, which is more closely related to which? Is A more closely related to B or to C? Figure 9.3 depicts what is known as a *cladogram*. That particular cladogram represents the hypothesis that A is more closely related to B than to C. Another way of putting the same point is to say that A shares a more recent common ancestor with B than it does with C. Or to put it still another way: A and B belong to a monophyletic group to which C does not belong. Different cladograms represent different hypotheses about past evolutionary relationships.

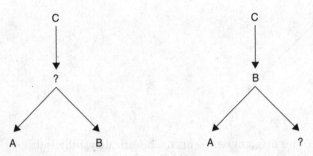

Figure 9.4 Phylogenetic trees. Both of these trees contain information about what evolved from what. Both are compatible with the cladogram in Figure 9.3.

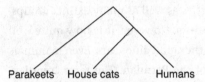

Parakeets   House cats   Humans

Figure 9.5 A cladogram known to be true. Humans share a more recent common ancestor with house cats than with parakeets.

A cladogram represents important information about evolutionary history, but it also leaves out a great deal. It says nothing about which species evolved from which. The cladogram in Figure 9.3 is compatible with several different stories about what evolved from what. It could be that *A* evolved from *B*, or that *B* evolved from *A*. Still another possibility is that *A* and *B* both evolved from some common ancestor. A phylogenetic tree is a rather different kind of diagram that conveys information about what evolved from what (see Figure 9.4). The main thing to appreciate is that cladists abstract away from some questions about evolution in order to focus more narrowly on others.

The only way to test a cladistic hypothesis is to look at similarities between organisms. The really crucial thing to see is that some similarities matter in cladistics while others don't. Cladists use the term *synapomorphy* for the similarities that matter, and the terms *homoplasy* and *symplesiomorphy* for the similarities that don't. Figure 9.5 depicts a cladogram that we know to be correct. We can point to some similarities that are shared by all three species: all have hearts; all have four limbs; etc., etc. These are all ancestral traits that they share because they all got them from some distant common ancestor (= *symplesiomorphies*, or shared ancestral traits). We have to just ignore these similarities because they contain no interesting information. Notice also that both humans and parakeets are bipedal; both like to eat seeds; both can

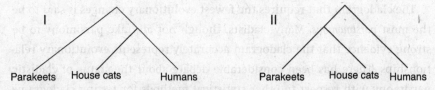

Figure 9.6 Two rival cladograms. Each represents a different hypothesis about the past.

talk, etc., but that cats have none of these traits. These similarities are called *homoplasies*, and because they arose independently over the course of evolution, they don't count as evidence; they don't tell us that humans and parakeets belong to a monophyletic group that cats don't belong to. The similarities that matter are the *synapomorphies*: both humans and cats have hair, both give birth to their young live, etc., etc. These are shared derived traits that both humans and house cats have inherited from a common mammalian ancestor.

How do scientists test cladistic hypotheses? Suppose we want to assess the two cladograms in Figure 9.6. Cladists think that the best way to approach this question is to ask: Which cladogram requires the simplest (= most parsimonious) evolutionary process? Suppose we pick out 100 traits, and suppose that all of the traits we're looking at come in two varieties, + or −. Suppose that at the start of the evolutionary process, all traits are set to +. And suppose the following:

Parakeets have 20 minuses.
Humans have 80 minuses.
House cats have 75 minuses.

Now it's easy to add up how many evolutionary changes each cladogram requires.

Cladogram I:	20 changes (to get to parakeets)
	75 changes (to get to the human/house cat group)
	5 changes (to get to humans)
	100 changes total
Cladogram II:	80 changes (to get to humans)
	20 changes (to get to the parakeet/house cat group)
	55 changes (to get to house cats)
	155 changes in all

The cladogram that requires the fewest evolutionary changes is said to be the most *parsimonious*. Many cladists, though not all, take parsimony to be strong evidence that the cladogram accurately represents evolutionary relationships. There has been considerable debate about the status of cladistic parsimony with respect to other statistical methods for testing cladograms (see Sober 1988 for an excellent introduction to the issues). All agree, however, that the assessment of cladistic hypotheses needs to be rigorous and quantitative. Because the number of possible cladograms describing evolutionary relationships among species increases dramatically with more than three species, analyzing cladograms for parsimony is a major computational challenge. These days it is usually done with the help of computer programs. In addition, it is becoming more and more common to use genetic information, rather than morphological characters, as the basis for phylogenetic reconstruction. The fact that cladistics permits rigorous empirical tests of historical hypotheses has led at least one writer to say that "without cladistics, paleontology is no more of a science than the one that proclaimed that the Earth was 6,000 years old and flat" (Gee 1999, p. 10).

As Brysse (2008) points out, paleontologists were a little slower to adopt the cladistic approach than other evolutionary biologists. She argues that when the change finally occurred, beginning in the 1980s, it led scientists to revisit the classification of the Burgess Shale fauna one more time. This time, they came armed with new concepts from cladistics that made it easier to explain the relationship between the weird animals of the Burgess Shale and existing groups. One of the most important new ideas from cladistics is the distinction between *stem groups* and *crown groups* (Jeffries 1979). Consider the interesting evolutionary scenario depicted in the cladogram in Figure 9.7. The total group consists of an early ancestor, plus all and only its descendants – a clear monophyletic group. However, *within* the total group, there is another monophyletic group, defined as the crown group. If you focus on only the species living today, it turns out that those species belong to two monophyletic groups – the crown group and the total group. The crown group is formed by making a cut just beyond the most recent common ancestor of the living species in group 1. The total group is formed by making a cut just above the common ancestor that the species in group 1 share with those in group 2. The stem group, which is entirely extinct, is what's left over once the crown group is subtracted from the total group. Notice that the stem group is not monophyletic, because

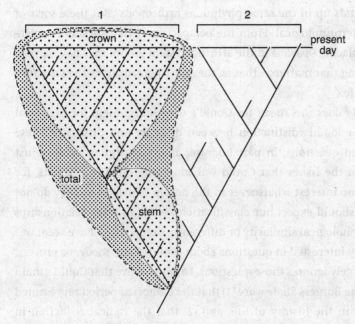

Figure 9.7 Stem group vs. crown group, based on Jeffries (1979), with permission from the Systematics Association. Note that the total group is the union of the stem group and the crown group.

there are many species that descended from the stem group and are not included in it.

Paleontologists Derek Briggs and Richard Fortey (1989) were the first to approach the Burgess Shale organisms with these distinctions in mind. They argued that instead of classifying creatures such as *Opabinia* in entirely new phyla, it makes more sense to place them in stem groups. Thus, if you think of arthropods as a crown group, you might place *Opabinia* in the extinct stem group. In a sense, *Opabinia* both does and does not belong in the same mono-phyletic group as arthropods. It is not a part of the arthropod crown group, but it is a part of the larger, total group. This is the cladistic solution to the Burgess Shale puzzle (Brysse 2008; MacLaurin and Sterelny 2008). The taxon-omy that Gould assumed in the writing of his 1989 book is, sadly, obsolete. There is still an issue about how to merge the old-fashioned talk of phyla with the new cladistic approach. Should we say that the crown group is a phylum, in which case *Opabinia* ends up in a different phylum from the true arthropods? Or should we say that the total group is the phylum, in which case *Opabinia* ends up in the same phylum as arthropods? But these sorts of questions are terminological. From the perspective of cladistics, once scientists agree on placing *Opabinia* in the arthropod stem group, they have agreed about everything that matters – that is, they have agreed about the structure of the tree of life.

What exactly does this mean for Gould's overall argument for historical contingency, or for his distinction between diversity and disparity? These are complicated questions, in part because, as Brysse argues, cladists just ignore many of the issues that Gould was most interested in. Cladists, for instance, take no interest whatsoever in the notion of disparity. They do not think that we should expect our classification system to track relationships of overall morphological similarity or difference. Gould, as we have seen, was also profoundly interested in questions about evolutionary mode, or process, but cladism largely ignores those questions. I argued above that Gould's main claims about the Burgess Shale were (1) that the Cambrian period represented peak disparity in the history of life, and (2) that the radical reduction in disparity that later occurred was unbiased. The reclassification of the Burgess Shale organisms does not really bear on either of these claims. The placing of *Opabinia* in an arthropod stem group is perfectly compatible with saying that *Opabinia* had a distinct body architecture, or *Bauplan*, that was quite different from that which characterizes the arthropod crown group. That body type was later lost for all time, and for all we know, that loss was simply a matter of chance – and our good fortune.

10   Molecular fossils

Molecular clocks *vs.* the fossil record

Evolutionary change involves certain changes at the molecular level: the replacement of some bits of DNA by others, as well as the replacement of some amino acids – the building blocks of proteins – by others. These days, it is a fairly straightforward matter for scientists to look at DNA sequences in different living organisms. It is also possible to measure the differences in DNA with great precision. To give just one example, Heckman et al. (2001) studied the proteins found in the cell nuclei of green algae, mosses, and plants. They measured differences in the amino acids that make up seventy-five different proteins found in the cell nuclei of each of these three groups. If you know the average rate of change in the genes or proteins, then you should be able to work backwards in order to determine how much evolutionary time it took, at that rate of change, to get from a common ancestor in the past to the living organisms that we observe today. Using this approach, Heckman and colleagues estimated that land plants diverged from green algae a very long time ago – just over 1 billion years, give or take 100 million or so. This is just one of many recent studies that take advantage of what has come to be known as the "molecular clock" in order to date past evolutionary events, and in particular those points of divergence where the tree of life splits.

Ayala (2009) offers the example of a single protein, cytochrome-c. All animals have versions of this protein that contain 104 amino acids, but not all animals have exactly the same combination of amino acids. For example, humans and rhesus monkeys have 103 identical amino acids and differ at only one place. Humans and horses have versions of cytochrome-c that are the same at ninety-two spots and differ at twelve. The larger difference between humans and horses suggests that our common ancestor with horses is much further back in the past. If we knew the rate of change, we could also draw conclusions about how far back in time those evolutionary divergences occurred.

In order to use the molecular clock to date past evolutionary events, scientists first need to know the average rate of nucleotide and/or amino acid replacement. (Nucleotides are the chemicals – adenine, cytosine, thymine, and guanine – that make up DNA.) Estimating the rate of change requires *calibration*. Suppose we know unequivocally from the fossil record that species *A* and species *B* diverged 20 million years ago. And suppose that *A* and *B* have versions of cytochrome-c that differ with respect to two amino acid positions. This imaginary example would suggest that differences in the amino acids arise once every 10 million years. By comparing a number of such examples, where the time of divergence is well known from the fossil record, it is possible to obtain an estimate of the rate of change in the amino acid sequence. Once in possession of that information about the rate of change, scientists can begin to date past evolutionary events based simply on measured differences in the composition of proteins and genes found in living organisms. Calibration of the molecular clock actually presupposes that the fossil record at least sometimes provides accurate dates for past evolutionary events (Benton 2009, p. 50).

Surprisingly, though, several studies using the molecular clock have produced dates that are wildly out of line with what the fossil record seems to say. Wray, Levinton, and Shapiro (1996) argued, based on the molecular clock, that the major groups of multicellular animals must have diverged as long ago as 1 billion years. Yet the oldest fossil remains of the major Metazoan groups date from the time of the Cambrian explosion, around 550–540 million years ago. In other words, the molecular clock seemed to say that the major Metazoan groups are twice as old as one would think based on the fossil record alone. Similarly wild discrepancies have shown up in other studies. A team of scientists led by Blair Hedges argued, again based on the molecular clock, that the major orders of mammals must have diverged around 120–130 million years ago (Hedges et al. 1996). Yet the oldest fossil mammals that can be placed clearly in any of the modern orders occur after the extinction of the dinosaurs 65 million years ago. Although there are much older mammal fossils from the time of the dinosaurs, they supply no clear evidence of the divergence of the modern orders – rodents, bats, carnivores, primates, and so on. I noted above that Heckman and colleagues found that land plants must have diverged from green algae around a billion years ago, at least according to the molecular clock. They argued that land plants diverged from all other plants somewhat more recently, around 700 million years ago. Yet there remains a 225-million-year gap between that date and the earliest appearance of land plants in the fossil record (Wellman 2004, p. 135). These discrepancies have generated a vigorous and rather messy debate.

Benton (2009) points out that different paleontologists reacted to these studies in different ways. One group, whom he calls the "ostriches," just "ignored the molecular challenge and hoped that it would go away" (2009, p. 46). Another group, the "lapdogs," assumed that the molecular clock must be telling the "true" evolutionary time. Notice that if you take that view, you are committed to saying that the fossil record is incomplete. If land plants really evolved 700 million years ago, then the fossil record contains no information about the first 225 million years of land plant evolution. This view is closely related to an old idea that the fossil record is strongly biased toward the present. The further back in the past you go, the less complete the record is. If that view is correct, it would help explain why the fossil record is a less accurate indicator of the timing of evolutionary events than the molecular clock. The third group of paleontologists, among whom Benton includes himself, are the "mules" who stubbornly insisted on the accuracy of the fossil record. The mules thought that there must be something wrong with the dates generated by the molecular studies.

To all appearances, the debate seems to pit paleontology against molecular biology. Each discipline has its own source of evidence – the fossil record *vs.* the genes and proteins of living creatures – and the issue is which of these sources of evidence can tell us more about the past. The relative importance of paleontology as a contributor to evolutionary science is one of the things at stake in this debate, for paleontology's disciplinary status and prestige have always been tied up with questions about the completeness of the fossil record. Darwin dealt an early blow to paleontology when, in the *Origin of Species*, he lamented the incompleteness of the geological record. Over a century later, Gould and Eldredge launched the paleobiological revolution by arguing that the fossil record is more complete than anyone had realized because the very gaps that Darwin complained about contain information. Now, at the height of the paleobiological revolution, when paleontologists have become virtuosos at documenting patterns in the fossil record and assessing claims about evolutionary processes, molecular biology raises all the old worries: What if the fossil record is so incomplete that it offers a radically misleading picture of evolutionary history? The whole debate about the molecular clock is a debate about how similar our own planet is to Afossilia.

Paleontologists have long wondered whether the fossil record might be strongly biased toward the present. The idea has a lot of intuitive plausibility: the further back in time you go, the less complete the fossil record gets, in part because geological processes have had more time to destroy the evidence. In one important early discussion of this issue, Raup (1972) suggested that this bias toward the present could create the illusion of biodiversity increase. Raup noted that estimates of biodiversity in the fossil record correlate roughly with the volume of sedimentary rock laid down in a given time interval. The more sedimentary rock, the more biodiversity shows up in the fossil record. More recent historical periods are associated with larger volumes of sedimentary rock. The apparent trend toward greater biodiversity might just be a matter of how much biodiversity was recorded in the rocks. Paleontologists continue to work on this problem (see, for instance, Peters and Foote 2001). The clash between the molecular data and the fossil record only exacerbated the worry that the fossil record is strongly biased toward the present.

As it happens, further work has succeeded in reducing the discrepancies between the fossil and the molecular data. The metaphor of the clock suggests that molecular-level changes occur at a more or less constant rate. If the rates of change varied in different lineages, that could mess up the estimates. Many of these studies concern evolutionary events that happened way, way back in the distant past, such as the divergence of deuterostomes (chordates and echinoderms) from other multicellular animals. What if a gene changes at a slower rate in these clades than in other groups? Another possibility is that rates of change could increase or decrease over time. It is obviously unwise simply to assume that the rate of molecular change has been constant over time. However, testing claims about the rates of change has proven to be a tricky business. As the techniques for carrying out these tests have improved, some of the molecular clock estimates of the timing of past branching episodes have fallen much closer into line with estimates based on the fossil record. Some discrepancies remain, but the widely differing dates that worried scientists in the 1990s have to some extent been reconciled (Benton 2009; Ayala 2009). Remaining discrepancies could well indicate places where the fossil record is incomplete. In a sense, the fossil record served as a helpful check on the early study of molecular clocks. It is now widely recognized that some of the methods used in those early studies led to overestimates (Rodriguez-Trelles, Tarrio, and Ayala 2003). The early angst about the possibility that the fossil record is radically incomplete has given way to a growing consensus that comparing the fossil data with the molecular data is just one of many ways of studying the incompleteness of the fossil record.

The evolution of "fossils"

Philosophers of science have long taken an interest in questions about the meaning of scientific terms. Philosophy of science came into its own as a discipline in the mid twentieth century, at a time when many philosophers had taken the "linguistic turn," and were intensely concerned with questions about meaning. Philosophers and historians of science have noted that the following two things seem to be true:

Meaning change. The meanings of technical terms in science change over time.

Theory-dependence of meaning. The meanings of scientific terms usually depend on scientific theories.

Thomas Kuhn argued that the meanings of scientific terms change radically every time there is a scientific revolution. For example, the term "mass" meant one thing to Newtonian physicists in the nineteenth century, and quite another thing after Einstein showed how mass and energy are related. "Temperature" had one meaning for scientists who accepted the caloric theory of heat, according to which heat is a fluid, and quite another for later scientists who defined temperature in terms of the mean kinetic energy of the molecules of a gas. The meanings of scientific terms can change even without a full-blown scientific revolution. Meaning change is a phenomenon that can sometimes be gradual, sometimes punctuated. The second of the two claims above helps explain the first. If the meanings of scientific terms depend on background theories, then the meanings will change whenever the theories do.

Historian of science M.J.S. Rudwick (1976) showed how the meaning of the term "fossil" has evolved over time. The original literal meaning of "fossil" is just "obtained by digging." Back in the sixteenth century, collectors used the term in a much broader way than we do today, to refer to anything interesting dug out of or found lying around on the ground. That could include gemstones and even artifacts as well as organic remains. Indeed, one of the major challenges for early students of fossils was to determine just what did and what didn't have organic origins.

Early European settlers in the Blue Ridge Mountains of Virginia and North Carolina were surprised to find small stony crosses when they cleared and tilled the land. We today know that these "fairy crosses" are merely geological phenomena. They are naturally occurring crystals made of staurolite, which contains iron, magnesium, and zinc. But a legend sprang up – perhaps among the early European settlers, perhaps among the Cherokee who already lived in the region. According to that legend, "little people" living in the woods had shed tears when they heard the news of Jesus's crucifixion. When those tears fell upon the ground, they took the shape of stony crosses. Are fairy crosses fossils? It depends. According to the sixteenth-century usage, they certainly would count as fossils, even if the legend were true. According to our narrower usage, according to which "fossil" refers only to organic remains, we would say that fairy crosses are not fossils, though someone who really believes the legend might say that they are.

The religious explanation of the occurrence of fairy crosses may sound simplistic, but it is a version of what was once the going scientific theory about the origins of fossils. Collectors knew that rock strata often contain beautiful facsimiles of plants and shellfish, as well as other living things, but no one knew for sure where those facsimiles came from. Did they somehow emerge spontaneously within the earth as a result of geological processes? Did they somehow come from living things? According to one early version of the latter view, fossils developed in the rocks when reproductive material – semen, seeds, and so forth – fell to the ground in the right sort of geological conditions. This reproductive material somehow communicated the form of the living creature to the rocks. Even once you narrow the use of the term "fossil" to refer only to organic remains, the meaning of the term depends largely on your theory about how fossils are formed. If we could travel back in time and converse with a scientist who held this reproductivist view, we and the scientist could point to the same item and say, "that's a fossil," while meaning very different things. We would mean that it is a product of a dead organism. The earlier scientist would mean that it is the product of a living one. The meaning of the term "fossil" depends on one's theory of fossilization.

The two claims I defended earlier – that the meanings of scientific terms change over time, and that they depend on theories – have the interesting consequence that people can easily end up talking past one another.

Creationists and Biblical literalists who deny evolution will occasionally say that God placed fossils in the rocks to test our faith that the Bible is the literal truth. However, if we define "fossil" in the way that paleontologists currently do – i.e., as the geological remains of living things – then it turns out that what the creationist says is false by definition. Because the creationist has such a radically different understanding of where fossils come from, the creationist and the evolutionist will be at risk of miscommunicating. Not that miscommunication is inevitable: surely each party can, with a little reflection, appreciate how the other side is using the term.

The meaning of the term "fossil" continues to evolve. For example, scientists sometimes talk of "living fossils," which are lineages that have persisted virtually unchanged for many millions of years. A good example is the Coelacanth, a type of fish that was known from Paleozoic rocks of the Devonian period, some 350 million years ago, around the time that vertebrates were first beginning to adapt to life on dry land. No one had thought that the Coelacanth had survived the K-T extinction event 65 million years ago, but in 1938 a Coelacanth was caught off the coast of South Africa (Weinberg 2000). Much more recently, scientists discovered large single-celled organisms living at the bottom of the Caribbean Sea. They are quite likely also living fossils from an even more distant Precambrian past, since they have left tracks in the sea floor that look just like tracks found in the fossil record as far back as the Precambrian, over a billion years ago (Matz et al. 2008). Is a living fossil *really* a fossil? Maybe not. Maybe the concept of a living fossil is merely a metaphorical one. What does it mean, though, to say that a concept is merely metaphorical? Or maybe the Coelacanth really is a fossil. We could take "fossil" to mean "any currently existing trace or remnant of prehistoric life," without requiring that fossils have anything to do with rocks. This definition implies that the genes and proteins in our own bodies are fossils. It might even imply that all living organisms are fossils.

Just what counts as a fossil is also subject to changes in technology. Over the last several years, scientists working in the emerging field of paleogenomics have rapidly improved techniques for sequencing the DNA of recently extinct organisms, including the giant cave bear (Noonan, Hofreiter, and Smith 2005), the woolly mammoth (Miller et al. 2008), the Tasmanian tiger (Miller et al. 2009), and even the Neanderthals (Green 2006). Scientists have even extracted bits of DNA from 80 million-year-old dinosaur bones, although no one hopes to be able to sequence an entire dinosaur genome (Woodward et al. 1994). In the case of the woolly mammoth, which persisted until 10,000–12,000 years ago, DNA samples are readily obtainable from the hair found on carcasses that remained frozen in Siberian and Alaskan permafrost for thousands of years. This molecular fossil evidence has helped to answer questions that would otherwise have remained a mystery. For example, we know that woolly mammoths roamed across Asia and North America during the Pleistocene. Were the North American and the Siberian populations genetically distinct, or did they represent one large interbreeding population with a huge geographical range?

Paleontologists use the term "trace fossil," or "ichnofossil," to refer to fossils that are not actually the remains of prehistoric life, but rather evidence of effects that prehistoric organisms had on their environments. Many trace fossils start out as imprints in mud or clay – think of dinosaur footprints. However, the concept of a trace fossil might be extended to include chemical signals in the geological record. To give just one example, we know that when different carbon isotopes are present, photosynthetic algae prefer some isotopes to others. As a result, green algae can have a global effect on the carbon isotope ratios in seawater. That, in turn, can show up as a chemical signal in carbonate rocks, which are formed on the ocean floor. The carbon isotope ratios in carbonate rocks represent a kind of chemical trace of the amount of photosynthetic activity in prehistoric oceans – a kind of chemofossil. Carbon isotope ratios do not, however, fall neatly under the traditional concept of a fossil, or even the traditional concept of an ichnofossil.

The study of chemofossils also promises to help scientists to date past evolutionary events. One of the major groups of Metazoans (multicellular animals) that shows up clearly in the fossil record of the Cambrian is the sponges. But when did the sponges first evolve? There is some fossil evidence for sponges living around 550 million years ago. However, Love and colleagues (2009) have pushed that date back by about 100 million years. They found "fossil steroids" in Precambrian rocks in Oman. When sponges die and decompose, their cell walls break down, leaving distinctive chemical traces, or biomarkers. Geochemical analysis reveals that the distinctive chemical signature of sponges can be found in sedimentary rocks as old as 635 million years, much older than the macrofossil record would suggest.

In one of the most dramatic recent developments in dinosaur science, paleontologist Mary Schweitzer led a team of scientists that successfully identified and analyzed protein fragments from a 68 million-year-old *Tyrannosaurus* skeleton (Schweitzer et al. 2007). They found that collagen from the walls of blood vessels can be preserved inside fossilized bone. Although many scientists were skeptical of the results at first, Schweitzer answered most of the skeptical doubts with a follow-up study of collagen preserved in the bones of an 80-million-year old hadrosaur (Schweitzer et al. 2009). This rapidly developing study of molecular dinosaur fossils may provide an entirely new line of evidence concerning evolutionary history by making it possible to compare the structures of dinosaur proteins for purposes of reconstruction of evolutionary relationships and dating evolutionary events.

To summarize: the meaning of the term "fossil" depends to a large extent on going scientific theories as well as on current technology, and it is subject to change. Although these observations may not seem too exciting at first, they have interesting consequences when you combine them with the observation that the fossil record has always been the disciplinary turf of paleontology, as well as the point made earlier that paleontology's status has waxed and waned with changing views about the completeness of the fossil record. What counts as the fossil record in the first place is up for scientific negotiation. The concept of the fossil record is itself a theoretical concept that scientists can broaden or narrow in ways that are informed by current theory and technology. This philosophical point about changes in the meanings of scientific terms is crucial to understanding the debate about the molecular clock.

It is possible to take a broader view of the fossil record, according to which the genes and proteins of living organisms would count as part of that record. Consider the following series of examples:

Teeth and bones from dinosaurs
Teeth and bones from more recently extinct cave bears
DNA obtained from the teeth and bones of cave bears
DNA obtained from the hair of frozen mammoth carcasses more than 10,000
    years old.
DNA obtained from the hair of stuffed specimens in museum collections, from
    species that went extinct over a century ago.
DNA obtained from living organisms.

It's not entirely clear where we should draw the line and say that we are not talking about fossils anymore. If you take the broadest possible view, then the work on the molecular clock, far from challenging the completeness of the fossil record, may well serve instead to expand our conception of what counts as a fossil. Such work may show that the fossil record is more complete than anyone had realized before. I hasten to add that my goal here is not to defend this broader conception of "fossil." The claim I am really interested in is the more modest one: if you construe the debate about the molecular clock as a debate about the completeness of the fossil record – and that is how many scientists construe it – then you are in effect working with a somewhat narrower conception of "fossil," and that narrower conception is optional.

In Chapter 1, I asked you to imagine a world without fossils, but what is a fossil? Without a clear definition of "fossil," it's not at all obvious how we should think about the imaginary planet of Afossilia. The relative completeness of the fossil record is partly an empirical issue, but it is partly also a conceptual issue. That is, it is partly a matter of how broadly or how narrowly we decide to construe the term "fossil." Afossilia represents the hypothetical limiting case of incompleteness in the fossil record, but it is also a test case for thinking about the meaning of the term "fossil." Would carbonate rocks on Afossilia contain carbon isotope ratios like those on Earth? Would the Afossilian tundra contain any frozen mammoth carcasses? Would there be any Coelacanths on Afossilia? How you envision Afossilia will depend on what you mean by "fossil" in the first place, and that in turn will make a difference to how much we want to say the Afossilians could know about evolution.

Empirical-conceptual knots: the proper focus for philosophy of science

In this book we have examined four major issues in evolutionary paleontology: PE, species selection and the hierarchical expansion of evolutionary theory, the study of directional evolutionary trends, and the debate about evolutionary contingency. One common theme has emerged from our exploration of these four issues. Conceptual questions and empirical questions in paleontology – and surely in natural science more broadly – are difficult to disentangle from one another. With respect to each of these four topics, we have begun with questions that seem like ordinary empirical questions of natural science and then worked our way upwards, pursuing bottom-up (or science-first) philosophy of science. In each case we ran into problems and questions that have both conceptual and empirical components.

First, in the discussion of PE, we found that although PE does involve an empirical claim about evolutionary tempos, it is also closely tied to the philosophical thesis that all observation is theory-laden. What's more, attempts to test PE against the fossil record raise some conceptual questions about species. The seemingly empirical claim that species undergo little morphological change once they become well established could simply be true by virtue of the way that paleontologists group fossil specimens into species – or so some critics of PE have argued. Next, in our discussion of species selection, we saw that the extent to which species selection really occurs in nature – what seems like an empirical issue – is closely related to the question of what species selection really requires. And that second question seems like a conceptual issue. Third, in the discussion of evolutionary trends, I presented an example of what philosophers of science call an underdetermination problem, or an empirical question that isn't answerable by the available evidence. The problem there was how to tell the difference between a directional trend that is created by a shifting bias and a trend that results from a constant bias plus a shifting upper boundary in the state space. But we saw that this underdetermination problem only arises as a result of the introduction of new concepts for the study of the dynamics of evolutionary trends. In this case, conceptual innovation leads to empirical underdetermination. Fourth, we found that the seemingly empirical question "Is evolution contingent?" is, in effect, several different questions rolled into one. There are several different senses of "contingency," and the answer to the empirical question may well depend on which of these senses one has in mind. And finally, we have seen just now that the question of the incompleteness of the fossil record, a question which is raised anew by recent work on molecular clocks, is partly a question about the meaning of "fossil."

I'd like to suggest, in closing, that these *empirical–conceptual knots* inevitably crop up sooner or later in the course of scientific research – I doubt that paleontology is exceptional in this regard. Moreover, these empirical–conceptual knots represent the appropriate sites of engagement between philosophy and the natural sciences. Many philosophers naturally gravitate toward these kinds of problems and questions, and they are most easily discovered by starting with the science and working from the bottom up. The knot imagery may not be perfect, but what I have in mind is something like this: imagine a tangled knot consisting of a variety of different colored strings. Let the different colors represent empirical *vs.* conceptual threads, or aspects of the problem. If you tug on one of the conceptual threads, that could either loosen or tighten the knot. In order to untie the knot, you might need to tug simultaneously on a conceptual one and an empirical one. If you tug in the wrong way, that just makes things worse. I don't claim to have untied any knots here, but I do hope that this book has made it plausible that philosophers of science have something to contribute to the investigation of empirical–conceptual knots. Philosophers, in particular, may have a special role to play in following the conceptual threads and seeing what happens when you tug on them. Not that philosophy is *merely* about concepts and definitions. The conceptual threads aren't all that interesting by themselves; they become interesting when they become tangled up with other empirical threads.

In this book I hope to have shown that evolutionary paleontology has its share of these conceptual–empirical knots. The very existence of such problems indicates that evolutionary paleontology deserves to be considered a serious theoretical science.

Bibliography

Alroy, J. 1998. "Cope's rule and the dynamics of body mass evolution in North American fossil mammals," *Science* 280: 731–734.

2000. "Understanding the dynamics of trends within evolving lineages," *Paleobiology* 26(3): 319–329.

Alvarez, L., Alvarez, W., Asaro, F., and Michel, H. 1980. "Extra-terrestrial cause for the Cretaceous–Tertiary extinction," *Science* 208: 1094–1108.

Arnold, A.J. and Fristrup, K. 1982. "The theory of evolution by natural selection: a hierarchical expansion," *Paleobiology* 8(2): 113–129.

Arnold, A.J., Kelly, D.C., and Parker, W.C. 1995. "Causality and Cope's rule: evidence from the planktonic foraminifera," *Journal of Paleontology* 69(2): 203–210.

Ayala, F.J. 1988. "Can 'progress' be defined as a biological concept?" in M.H. Nitecki (ed.) *Evolutionary Progress*. University of Chicago Press, pp. 75–96.

1998. "Beyond Darwinism? The challenge of macroevolution to the synthetic theory of evolution," in M. Ruse (ed.) *The Philosophy of Biology*. Amherst, NY: Prometheus Books, pp. 118–133.

2009. "Molecular evolution vis-à-vis paleontology," in D. Sepkoski and M. Ruse (eds.) *The Paleobiological Revolution: new essays on the growth of modern paleontology*. University of Chicago Press, pp. 176–198.

Barbour, I. 1997. *Religion and Science: historical and contemporary issues*. New York: HarperCollins.

Baron, C. 2009. "Epistemic values in the Burgess Shale debate," *Studies in History and Philosophy of Biology and Biomedical Sciences* 40: 286–295.

Beatty, J. 1995. "The evolutionary contingency thesis," in G. Wolters and J.G. Lennox (eds.) *Concepts, Theories, and Rationality in the Biological Sciences*. University of Pittsburgh Press, pp. 45–81.

1997. "Why do biologists argue like they do?" *Philosophy of Science* 64: S432–443.

2006. "Replaying life's tape," *The Journal of Philosophy* 103(7): 336–362.

Bedau, M. 1997. "Weak emergence," in J. Tomberlin (ed.) *Philosophical Perspectives 11: mind, causation, and world*. Cambridge, MA: Blackwell, pp. 375–399.

209

Bell, G.L., Jr. 1997. "Introduction [to the Mosasauridae]," in J.M. Callaway and E.L. Nicholls (eds.) *Ancient Marine Reptiles*. New York: Academic Press, 281–292.

Ben-Menahem, Y. 1997. "Historical contingency," *Ratio* 10: 99–107.

Benton, M.J. 2004. "The quality of the fossil record," in P.C.J. Donoghue and M.P. Smith (eds.) *Telling the Evolutionary Time: molecular clocks and the fossil record*. London: Taylor and Francis, pp. 66–90.

    2009. "The fossil record: biological or geological signal?" in D. Sepkoski and M. Ruse (eds.) *The Paleobiological Revolution: new essays on the growth of modern paleontology*. University of Chicago Press, pp. 42–59.

Benton, M.J., Wills, M.A., and Hitchin, R. 2000. "Quality of the fossil record through time," *Nature* 403: 534–537.

Brandon, R. 1997. "Does biology have laws? The experimental evidence," *Philosophy of Science* 64(4): S444–S457.

Brayard A., Nützel A., Stephen D.A., Bylund K.G., Jenks J., and Bucher H. 2010. "Gastropod evidence against the early Triassic Lilliput effect," *Geology* 38(2): 147–150.

Briggs, D.E.G. and Fortey, R. 1989. "The early radiation and relationship of the major arthropod groups," *Science* 246(4927): 241–243.

Brysse, K. 2008. "From weird wonders to stem lineages: the second reclassification of the Burgess Shale fauna," *Studies in History and Philosophy of Biology and Biomedical Sciences* 39: 298–313.

Butler, R.J. and Goswami, A. 2008. "Body size evolution in Mesozoic birds: little evidence for Cope's rule," *Journal of Evolutionary Biology* 21(6): 1673–1682.

Clayton, N.S. and Emery, N.J. 2008. "Canny corvids and political primates: a case for convergent evolution in intelligence," in S. Conway Morris (ed.) *The Deep Structure of Biology: is convergence sufficiently ubiquitous to give a directional signal?* West Conshohocken, PA: Templeton Foundation Press, pp. 128–142.

Cleland, C. 2001. "Historical science, experimental science, and the scientific method," *Geology* 29(11): 987–990.

    2002. "Methodological and epistemic differences between historical science and experimental science," *Philosophy of Science* 69(3): 474–496.

Conway Morris, S. 2003. *Life's Solution: inevitable humans in a lonely universe*. Cambridge University Press.

    (ed.) 2008a. *The Deep Structure of Biology: is convergence sufficiently ubiquitous to give a directional signal?* West Conshohocken, PA: Templeton Foundation Press.

    2008b. "Evolution and convergence: some wider considerations," in S. Conway Morris (ed.) *The Deep Structure of Biology: is convergence sufficiently ubiquitous to give a directional signal?* West Conshohocken, PA: Templeton Foundation Press, pp. 46–67.

Cope, E.D. 1974. *The Primary Factors of Organic Evolution*. Chicago, IL: Open Court Publishing Company.

Daeschler, E.B., Shubin, N.H., and Jenkins, Jr., F.A. 2006. "A Devonian tetrapod-like fish and the evolution of the tetrapod body plan," *Nature* 440: 757–763.

Darwin, C. 1859/1964. *On the Origin of Species*. Cambridge, MA: Harvard University Press.

Dawkins, R. 1976. *The Selfish Gene*. Oxford University Press.

1986. *The Blind Watchmaker*. New York: W.W. Norton.

Dennett, D. 1991. "Real patterns," *The Journal of Philosophy* 88(1): 27–51.

1995. *Darwin's Dangerous Idea: evolution and the meanings of life*. New York: Simon and Schuster.

Dommergues, J.-L., Montuire, S., and Neige, P. 2002. "Size patterns through time: the case of the early Jurassic ammonite radiation," *Paleobiology* 28(4): 423–424.

Donoghue, P.C.J. and Smith, M.P. (eds.) 2004. *Telling the Evolutionary Time: molecular clocks and the fossil record*. London: Taylor and Francis.

Duhem, P. 1991. *The Aim and Structure of Physical Theory*. Princeton University Press.

Dusek, V. 2003. "Steve Gould: Marxist as biologist," *Rethinking Marxism* 15(4): 451–466.

Eldredge, N. 1971. "The allopatric model and phylogeny in Paleozoic invertebrates," *Evolution* 25: 1228–1232.

1996. "Hierarchies in macroevolution," in Jablonski, D., Erwin, D.H., and Lipps, J.H. (eds.) *Evolutionary Paleobiology*. University of Chicago Press, pp. 42–61.

Eldredge, N. and Gould, S.J. 1972. "Punctuated equilibria: an alternative to phyletic gradualism," in T.J.M. Schopf (ed.) *Models in Paleobiology*. San Francisco, CA: Freeman, Cooper, & Co., pp. 85–115.

Elgin, M. 2006. "There may be strict empirical laws in biology, after all," *Biology and Philosophy* 21(1): 119–134.

Friis, E.M., Raunsgard Pedersen, K., and Crane, P.R. 2006. "Cretaceous angiosperm flowers: innovation and evolution in plant reproduction," *Palaeogeography, Palaeoclimatology, Palaeoecology* 232(2–4): 251–293.

Frohlich, M.W. and Chase, M.W. 2007. "After a dozen years of progress the origin of angiosperms is still a great mystery," *Nature* 450(20): 1184–1189.

Gallie, W.B. 1959. "Explanations in history and the genetic sciences," in P. Gardiner (ed.) *Theories of History*. Glencoe, IL: The Free Press.

1964. *Philosophy and the Historical Understanding*. New York: Schocken.

Gee, H. 1999. *In Search of Deep Time: beyond the fossil record to a new history of life*. Ithaca, NY: Cornell University Press.

Ghiselin, M. 1974. "A radical solution to the species problem," *Systematic Zoology* 23: 536–544.

Gingerich, P.D. 1984. "Punctuated equilibria – where is the evidence?" *Systematic Zoology* 33(3): 335–338.

Goldschmidt, R. 1940. *The Material Basis of Evolution*. New Haven, CT: Yale University Press.

Gould, G.C. and MacFadden, B.J. 2004. "Gigantism, dwarfism, and Cope's rule: 'Nothing in evolution makes sense without a phylogeny'," *Bulletin of the American Museum of Natural History* 285: 219–237.

Gould, S.J. 1977. "Return of the hopeful monster," *Natural History* 86: 22–30.

1980a. "Is a new and general theory of evolution emerging?" *Paleobiology* 6(1): 119–130.

1980b. "The promise of paleontology as a nomothetic, evolutionary discipline," *Paleobiology* 6(1): 96–118.

1988a. "On replacing the idea of progress with an operational notion of directionality," in M.H. Nitecki (ed.) *Evolutionary Progress*. University of Chicago Press, pp. 319–338.

1988b. "Trends as changes in variance: a new slant on progress and directionality in evolution," *Journal of Paleontology* 62: 319–329.

1989. *Wonderful Life: the Burgess Shale and the nature of history*. New York: W.W. Norton.

1993a. "The wheel of fortune and the wedges of progress," in Gould, S.J., *Eight Little Piggies: reflections in natural history*. New York: W.W. Norton, pp. 300–312.

1993b. "Betting on chance – and no fair peeking," in Gould, S.J., *Eight Little Piggies: reflections in natural history*. New York: W.W. Norton, pp. 396–408.

1996. *Full House: the spread of excellence from Plato to Darwin*. New York: W.W. Norton.

1997a. "Cope's rule as psychological artefact," *Nature* 385(6613): 199–200.

1997b. "Nonoverlapping magisteria," *Natural History* 106: 16–22.

2002. *The Structure of Evolutionary Theory*. Cambridge, MA: The Belknap Press of Harvard University Press.

Gould, S.J. and Eldredge, N. 1977. "Punctuated equilibria: the tempo and mode of evolution reconsidered," *Paleobiology* 3(2): 115–151.

1986. "Punctuated equilibrium at the third stage," *Systematic Zoology* 35(1): 143–148.

1993. "Punctuated equilibrium comes of age," *Nature* 366: 223–227.

Gould, S.J. and Lewontin, R. 1979. "The spandrels of San Marco and the panglossian paradigm: a critique of the adaptationist programme," *Proceedings of the Royal Society B* 205: 581–598.

Gould, S.J. and Lloyd, E.A. 1999. "Individuality and adaptation across levels of selection: how shall we name and generalize the unit of Darwinism?" *Proceedings of the National Academy of Sciences* 96: 11904–11909.

Gould, S.J. and Woodruff, D.S. 1990. "History as a cause of area effects: an illustration from Cerion on Great Inagua, Bahamas," *British Journal of the Linnean Society* 40: 67–98.

Grantham, T.A. 1995. "Hierarchical approaches to macroevolution: recent work on species selection and the 'effect hypothesis'," *Annual Review of Ecology and Systematics* 26: 301–321.

1999. "Explanatory pluralism in paleobiology," *Philosophy of Science* 66(supp): S223–S236.

2002. "Species selection," in M. Pagel (ed.), *Encyclopedia of Evolution*. Oxford University Press, pp. 1086–1087.

2007. "Is macroevolution more than successive rounds of microevolution?" *Palaeontology* 50(1): 75–85.

Green, R.E. 2006. "Analysis of one million base pairs of Neanderthal DNA," *Nature* 444: 330–336.

Hacking, I. 1999. *The Social Construction of What?* Cambridge, MA: Harvard University Press.

Hanson, N.R. 1958. *Patterns of Discovery: an inquiry into the conceptual foundations of science*. Cambridge University Press.

Heckman, D.S., et al. 2001. "Molecular evidence for the early colonization of land by fungi and plants," *Science* 293: 1129–1133.

Hedges, S.B., Parker, P.H., Sibley, C.G., and Kumar, S. 1996. "Continental breakup and the ordinal diversification of birds and mammals," *Nature* 381: 226–229.

Hempel, C.G. 1966. *Philosophy of Natural Science*. Englewood Cliffs, NJ: Prentice Hall.

Hone, D.W.E. and Benton, M.J. 2005. "The evolution of large size: how does Cope's rule work?" *Trends in Ecology and Evolution* 20(1): 4–6.

2007. "Cope's rule in the Pterosauria, and differing perceptions of Cope's rule at different taxonomic levels," *Journal of Evolutionary Biology* 20: 1164–1170.

Hone, D.W.E., Keesey, T.M., Pisanis, D., and Purvis, A. 2005. "Macroevolutionary trends in the Dinosauria: Cope's rule," *Journal of Evolutionary Biology* 18: 587–595.

Hone, D.W.E., Dyke, G.J., Haden, M., and Benton, M.J. 2008. "Body size evolution in Mesozoic birds," *Journal of Evolutionary Biology* 21(2): 618–624.

Horner, J.R. and Makela, R. 1979. "Nest of juveniles provides evidence of family structure among dinosaurs," *Nature* 282: 296–298.

Hull, D.L. 1988. *Science as a Process*. University of Chicago Press.

Hunt, G. and Roy, K. 2006. "Climate change, body size evolution, and Cope's rule in deep-sea ostracodes," *Proceedings of the National Academy of Sciences* 103(5): 1347–1352.

Huss, J. 2009. "The shape of evolution: the MBL model and clade shape," in D. Sepkoski and M. Ruse (eds.) *The Paleobiological Revolution: essays on the growth of modern paleontology*. University of Chicago Press, pp. 326–345.

Jablonski, D. 1987. "Heritability at the species level: analysis of geographic ranges of Cretaceous mollusks," *Science* 238: 360–363.

1996. "Body size and macroevolution," in D. Jablonski, D.H. Erwin, and J.H. Lipps (eds.) *Evolutionary Paleobiology*. University of Chicago Press, pp. 256–289.

1997. "Body-size evolution in Cretaceous mollusks and the status of Cope's rule," *Nature* 385: 250–252.

2008. "Species selection: theory and data," *Annual Review of Ecology and Systematics* 39: 501–524.

Jaffe, M. 2001. *The Gilded Dinosaur: the fossil wars between E.D. Cope and O.C. Marsh and the rise of American science*. New York: Three Rivers Press.

Jeffares, B. 2008. "Testing times: regularities in the historical sciences," *Studies in History and Philosophy of Biology and Biomedical Sciences* 39C: 469–475.

Jeffries, R.P.S. 1979. "The origin of chordates: a methodological essay," in M.R. House (ed.) *The Origin of Major Invertebrate Groups*. London: Academic Press, pp. 443–477.

Kemp. T.S. 1999. *Fossils and Evolution*. Oxford University Press.

Kimura, M. 1983. *The Neutral Theory of Molecular Evolution*. Cambridge University Press.

Kingsolver, J.G. and Pfennig, D.W. 2004. "Individual-level selection as a cause of Cope's rule of phyletic size increase," *Evolution* 58: 1608–1612.

Kitcher, P. 1984. "Species," *Philosophy of Science* 51: 308–333.

Kleinhans, M.G., Buskes, C.J.J., and de Regt, H. 2005. "Terra incognita: explanation and reductionism in Earth science," *International Studies in Philosophy of Science*, 19(3): 289–317.

Knouft, J.H. and Page, L.M. 2003. "The evolution of body size in extant groups of North American freshwater fishes: speciation, size distribution, and Cope's rule," *American Naturalist* 161(3): 413–421.

Kuhn, T.S. 1962/1996. *The Structure of Scientific Revolutions*. University of Chicago Press.

Laudan, L. and Leplin, J. 1991. "Empirical equivalence and underdetermination," *Journal of Philosophy* 88: 449–472.

Laurin, M. 2004. "The evolution of body size, Cope's rule, and the origin of amniotes," *Systematic Biology* 53(4): 594–622.

Lenski, R.E. and Travisano, M. 1994. "Dynamics of adaptation and diversification: a 10,000-generation experiment with bacterial populations," *Proceedings of the National Academy of Sciences* 91: 6808–6814.

Leplin, J. 1997. *A Novel Defense of Scientific Realism*. Oxford University Press.

Levinton, J.S. and Simon, C.M. 1980. "A critique of the punctuated equilibria model and implications for the detection of speciation in the fossil record," *Systematic Zoology* 29(2): 130–142.

Lewontin, R.C. 1970. "The units of selection," *Annual Review of Ecology and Systematics* 1: 1–18.

Lieberman, B.S. and Vrba, E. 2005. "Stephen Jay Gould on species selection: 30 years of insight," *Paleobiology* 31(2): 113–121.

Lloyd, E. and Gould, S.J. 1993. "Species selection on variability," *Proceedings of the National Academy of Sciences* 90: 595–599.

Love, G.D., et al. 2009. "Fossil steroids record the appearance of Demospongiae during the Cryogenian period," *Nature* 457: 718–721.

MacFadden, B.J. 1986. "Fossil horses from 'Eohippus' (Hyracotherium) to Equus: scaling, Cope's law, and the evolution of body size," *Paleobiology* 12(4): 355–369.

Machamer, P., Darden, L., and Craver, C. 2000. "Thinking about mechanisms," *Philosophy of Science* 57: 1–25.

MacLaurin, J. and Sterelny, K. 2008. *What Is Biodiversity?* University of Chicago Press.

Mallet, J. 2008. "Hybridization, ecological races, and the nature of species: empirical evidence for the ease of speciation," *Philosophical Transactions of the Royal Society B* 263: 2971–2986.

Malmgren, B.A., Berggren, W.A., and Lohmann, G.P. 1983. "Evidence for punctuated gradualism in the late Neogene Globorotalia tumida lineage of planktonic foraminifera," *Paleobiology* 9(4): 377–389.

Massare, J.A. 1988. "Swimming capabilities of Mesozoic marine reptiles: implications for method of predation," *Paleobiology* 14(2): 187–205.

Matz, M.V., Frank, T.M., Marshall, N.J., Widder, E.A., and Johnsen, S. 2008. "Giant deep-sea protest produces bilaterian-like traces," *Current Biology* 18: 1849–1854.

Maynard Smith, J. and Szathmáry, E. 1995. *The Major Transitions in Evolution*. New York: W.H. Freeman.

Mayr, E. 1942. *Systematics and the Origin of Species*. New York: Columbia University Press.

  1963. *Animal Species and Evolution*. Cambridge, MA: Harvard University Press.

  1988. *Toward a New Philosophy of Biology*. Cambridge, MA: Harvard University Press.

  1992. "Speciational evolution or punctuated equilibria," in I.A. Somit and S. Peterson (eds.) *The Dynamics of Evolution*. New York: Cornell University Press, pp. 21–48.

McKinney, M.L. 1990. "Trends in body-size evolution," in K.J. McNamara (ed.) *Evolutionary Trends*. Tucson, AZ: University of Arizona Press, pp. 75–120.

McNamara, K.J. (ed.) 1990. *Evolutionary Trends*. Tucson, AZ: University of Arizona Press.

McShea, D.W. 1991. "Complexity and evolution: what everybody knows," *Biology and Philosophy* 6: 303–324.

1994. "Mechanisms of large-scale evolutionary trends," *Evolution* 48: 1747–1763.

1996. "Metazoan complexity and evolution: is there a trend?" *Evolution* 50(2): 477–492.

1998. "Possible largest-scale trends in organismal evolution: eight 'live hypotheses'," *Annual Review of Ecology and Systematics* 29: 293–318.

2005. "The evolution of complexity without natural selection: a possible large-scale trend of the fourth kind," *Paleobiology* 31(supp): 146–156.

Mellor, D.H. 2005. *Probability: a philosophical introduction*. London: Routledge.

Miller, W., et al. 2008. "Sequencing the nuclear genome of the extinct woolly mammoth," *Nature* 456(20): 387–390.

Miller, W., Drautz, D.I., Janecka, J.E., et al. 2009. "The mitochondrial genome sequence of the Tasmanian tiger (*Thylacinus cynocephalus*)," *Genome Research* 19: 213–220.

Millstein, R.L. 2000. "Chance and macroevolution," *Philosophy of Science* 67(4): 603–624.

Mitchell, S. 2003. *Biological Complexity and Integrative Pluralism*. Cambridge University Press.

Moore, G.E. 1903/2004. *Principia Ethica*. New York: Dover Publications.

Newell, N.D. 1949. "Phyletic size increase, an important trend illustrated by fossil invertebrates," *Evolution* 3(2): 103–124.

Niklas, K.J. 1997. "Adaptive walks through fitness landscapes for early vascular land plants," *American Journal of Botany* 84(1): 16–25.

1998. "Evolutionary walks through a land plant morphospace," *Journal of Experimental Botany* 50(330): 39–52.

Nitecki, M.H., (ed.) 1988. *Evolutionary Progress*. University of Chicago Press.

Noonan, J.P.M., Hofreiter, D., Smith, D., et al. 2005. "Genomic sequencing of Pleistocene cave bears," *Science* 309(5734): 597–600.

Okasha, S. 2006. *Evolution and the Levels of Selection*. Oxford University Press.

Osborn, H.F. 1899. "A complete mosasaur skeleton, osseous and cartilaginous," *Memoirs of the American Museum of Natural History* 1(4): 167–188.

Parsons, K.M. 2001. *Drawing Out Leviathan: dinosaurs and the science wars*. Bloomington, IN: Indiana University Press.

Patterson, C. and Smith, A.B. 1987. "Is periodicity of mass extinctions a taxonomic artifact?" *Nature* 330: 248–251.

1989. "Periodicity in extinction: the role of systematics," *Ecology* 70: 802–811.

Peirce, C.S. 1955. *The Philosophical Writings of Peirce*, J. Buchler (ed.) New York: Dover Publications.

Peters, S.E. and Foote, M. 2001. "Biodiversity in the Phanerozoic: a reinterpretation," *Paleobiology* 27: 583–601.

Polly, P.D. 1998. "Cope's rule," *Science* 282: 51.

Popper, K.R. 1979. *Objective Knowledge: an evolutionary approach*, Second Edition. Oxford University Press.

    1996. "Darwinism as a metaphysical research program," in M. Ruse (ed.) *But Is It Science?* Amherst, NY: Prometheus Books, pp. 144–155.

Post, J.F. 1991. *Metaphysics: a contemporary introduction*. New York: Paragon House.

Princehouse, P. 2009. "Punctuated equilibrium and speciation: what does it mean to be a Darwinian?" in D. Sepkoski and M. Ruse (eds.) *The Paleobiological Revolution: essays on the growth of modern paleontology*. University of Chicago Press, pp. 149–175.

Prothero, D.R. 1992. "Punctuated equilibrium at twenty: a paleontological perspective," *Skeptic* 1(3): 38–47.

Quine, W.V. 1951. "Two dogmas of empiricism," *Philosophical Review* 60(1): 20–43. Reprinted in Quine, W.V. 1980. *From a Logical Point of View*. Cambridge, MA: Harvard University Press.

Raup, D.M. 1966. "Geometric analysis of shell coiling: general problems," *Journal of Paleontology* 40: 1178–1190.

    1972. "Taxonomic diversity during the Phanerozoic," *Science* 177(4054): 1065–1071.

    1985. *The Nemesis Affair: a story of the death of dinosaurs and the ways of science*. New York: W.W. Norton.

    1988. "Testing the fossil record for evolutionary progress," in M.H. Nitecki (ed.) *Evolutionary Progress*. University of Chicago Press, pp. 293–317.

    1991. *Extinction: bad genes or bad luck?* New York: W.W. Norton.

Raup, D.M. and Michelson, A. 1965. "Theoretical morphology of the coiled shell," *Science* 147: 1294–1295.

Raup, D.M. and Gould, S.J. 1974. "Stochastic simulation and the evolution of morphology – towards a nomothetic paleontology," *Systematic Zoology* 23: 305–322.

Raup, D.M., Gould, S.J., Schopf, T.J.M., and Simberloff, D. 1973. "Stochastic models of phylogeny and the evolution of diversity," *Journal of Geology* 81: 525–542.

Raup, D.M. and Sepkoski, J.J. 1984. "Periodicity of extinctions in the geologic past," *Proceedings of the National Academy of Sciences* 81(3): 801–805.

Roach, J. 2006. "Grizzly–polar bear hybrid found – but what does it mean?" *National Geographic News*. Available online at http://news.nationalgeographic.com/news/2006/05/polar-bears.html. Last accessed February 10, 2010.

Rodriguez-Trelles, F., Tarrio, R., and Ayala, F.J. 2003. "Molecular clocks: whence and whither?" in P.C.J. Donoghue and M.P. Smith (eds.) *Telling the Evolutionary Time: molecular clocks and the fossil record*. London: Taylor and Francis, pp. 5–26.

Rosenberg, A. and McShea, D.W. 2007. *Philosophy of Biology: a contemporary introduction*. London: Routledge.

Rudwick, M.J.S. 1976. *The Meaning of Fossils: episodes in the history of palaeontology*. University of Chicago Press.

Ruse, M. 1992. "Biological species: natural kinds, individuals, or what?" in M. Ereshefsky (ed.) *The Units of Evolution: essays on the nature of species*. Cambridge, MA: MIT Press, pp. 343–362.

1996. *Monad to Man: the concept of progress in evolutionary biology*. Cambridge, MA: Harvard University Press.

1999. *Mystery of Mysteries: is evolution a social construction?* Cambridge, MA: Harvard University Press.

2004. *Can a Darwinian be a Christian?* Cambridge University Press.

2009. "Punctuations and paradigms: has paleobiology been through a paradigm shift?" in D. Sepkoski and M. Ruse (eds.) *The Paleobiological Revolution: essays on the growth of modern paleontology*. University of Chicago Press, pp. 518–528.

Schmidt, D.N., Thierstein, H.R., and Bollman, J. 2004. "The evolutionary history of size variation in planktonic foraminiferal assemblages in the Cenozoic," *Palaeogeography, Palaeoclimatology, Palaeoecology* 212: 159–180.

Schopf, T.J.M. 1979. "Evolving paleontological views on deterministic and stochastic approaches," *Paleobiology* 5(3): 337–352.

Schweitzer, M.H., Suo, Z., Avci, R., et al. 2007. "Analyses of soft tissue from Tyrannosaurus rex suggest the presence of protein," *Science* 316: 277–280.

Schweitzer, M.H., Zheng, W., Organ, C.L., et al. 2009. "Biomolecular characterization and protein sequences of the Campanian Hadrosaur B. canadensis," *Science* 324: 626–631.

Sepkoski, D. 2005. "Stephen Jay Gould, Jack Sepkoski, and the 'quantitative revolution' in American paleobiology," *Journal of the History of Biology* 38: 209–237.

2009a. "The emergence of paleobiology," in D. Sepkoski and M. Ruse (eds.) *The Paleobiological Revolution: essays on the growth of modern paleontology*. University of Chicago Press, pp. 15–42.

2009b. "'Radical' or 'conservative'? The origin and early reception of punctuated equilibrium," in D. Sepkoski and M. Ruse (eds.) *The Paleobiological Revolution: essays on the growth of modern paleontology*. University of Chicago Press, pp. 301–325.

Sepkoski, D. and Raup, D.M. 2009. "An interview with David M. Raup," in D. Sepkoski and M. Ruse (eds.) *The Paleobiological Revolution: essays on the growth of modern
paleontology*. University of Chicago Press, pp. 459–470.

Sepkoski, D. and Ruse, M. (eds.) 2009. *The Paleobiological Revolution: essays on the growth
of modern paleontology*. University of Chicago Press.

Sepkoski, J.J., Jr. 1978. "A kinetic model of Phanerozoic taxonomic diversity. I.
Analysis of marine orders," *Paleobiology* 4: 223–251.

1979. "A kinetic model of Phanerozoic taxonomic diversity. II. Early Paleozoic
families and multiple equilibria," *Paleobiology* 5: 222–252.

1984. "A kinetic model of Phanerozoic taxonomic diversity. III. Post-paleozoic
families and mass extinctions," *Paleobiology* 10: 246–267.

Shanahan, T. 2004. *The Evolution of Darwinism: selection, adaptation, and progress in
evolutionary biology*. Cambridge University Press.

Simpson, C. and Harnik, P.G. 2009. "Assessing the role of abundance in marine
bivalve extinction over the post-Paleozoic," *Paleobiology* 35(4): 631–647.

Simpson, G.G. 1944. *Tempo and Mode in Evolution*. New York: Columbia University
Press.

1949. *The Meaning of Evolution*. New Haven, CT: Yale University Press.

Sober, E. 1988. *Reconstructing the Past: parsimony, evolution, and inference*. Cambridge,
MA: MIT Press.

1994. "Progress and direction in evolution," in J.H. Campbell and J.W. Schopf
(eds.) *Creative Evolution?!* Boston, MA: Jones and Bartlett Publishers.

1997. "Two outbreaks of lawlessness in recent philosophy of biology," *Philosophy
of Science* 64(4): S458–S467.

2000. *Philosophy of Biology*, Second Edition. Boulder, CO: Westview Press.

Stanford, P.K. 2001. "Refusing the Devil's bargain: what kind of underdetermination should we take seriously?" *Philosophy of Science* 68(supp):
S1–S12.

Stanley, S.M. 1973. "An explanation for Cope's rule," *Evolution* 27(1): 1–26.

1975. "A theory of evolution above the species level," *Proceedings of the National
Academy of Sciences* 72(2): 646–650.

1979. *Macroevolution: pattern and process*. San Francisco, CA: W.H. Freeman and
Company.

Sterelny, K. 1992. "Punctuated equilibrium and macroevolution," I.P.E. Griffiths (ed.) *Trees of Life: essays in the philosophy of biology*. Dordrecht: Kluwer,
pp. 41–64.

2001. *Dawkins vs. Gould: survival of the fittest*. Cambridge: Icon Books.

2005. "Another view of life," *Studies in History and Philosophy of Biology and Biomedical
Sciences* 36: 585–593.

2007. "Macroevolution, minimalism, and the radiation of the animals," in D.L. Hull and M. Ruse (eds.) *The Cambridge Companion to the Philosophy of Biology*. Cambridge University Press.

Sterelny, K. and Griffiths, P.E. 1999. *Sex and Death: an introduction to philosophy of biology*. University of Chicago Press.

Thiessen, G. 2006. "The proper place of hopeful monsters in evolutionary biology," *Theory in Biosciences* 124: 199–212.

2009. "Saltational evolution: hopeful monsters are here to stay," *Theory in Biosciences* 128: 43–51.

Travisano, M., Mangold, J.A., Bennett, A.F., and Lenski, R.E. 1995. "Experimental tests of the roles of adaptation, chance, and history in evolution," *Science* 27(5194): 87–90.

Tucker, A. 2004. *Our Knowledge of the Past*. Cambridge University Press.

Turner, D.D. 2000. "The functions of fossils: inference and explanation in functional morphology," *Studies in History and Philosophy of Science C: biology and biomedical sciences* 31: 193–212.

2005. "Local underdetermination in historical science," *Philosophy of Science* 72: 209–230.

2007. *Making Prehistory: historical science and the scientific realism debate*. Cambridge University Press.

2009a. "How much can we know about the causes of evolutionary trends?" *Biology and Philosophy* 24: 341–357.

2009b. "Beyond detective work: empirical testing in paleobiology," in D. Sepkoski and M. Ruse (eds.) *The Paleobiological Revolution: essays on the growth of modern paleontology*. University of Chicago Press, pp. 201–214.

van Valkenbergh, B., Wang, X., and Damuth, J. 2004. "Cope's rule, hypercarnivory, and extinction in North American canids," *Science* 306: 101–104.

Vrba, E. 1983. "Macroevolutionary trends: new perspectives on the roles of adaptation and incidental effect," *Science* 221(4608): 387–389.

1984. "What is species selection?" *Systematic Zoology* 33: 318–328.

1987. "Ecology in relation to speciation rates: some case histories of Miocene-Recent mammal clades," *Evolutionary Ecology* 1: 283–300.

1989. "Levels of selection and sorting, with special reference to the species level," *Oxford Surveys of Evolutionary Biology* 6: 111–168.

Vrba, E. and Eldredge, N. 1984. "Individuals, hierarchies, and processes: towards a more complete evolutionary theory," *Paleobiology* 10: 146–171.

Vrba, E. and Gould, S.J. 1986. "The hierarchical expansion of sorting and selection: sorting and selection cannot be equated," *Paleobiology* 12(2): 217–228.

Wagner, P.J. 1996. "Contrasting the underlying patterns of active trends in morphologic evolution," *Evolution* 50(3): 990–1007.

Wang, S.C. 2001. "Quantifying passive and driven large-scale evolutionary trends," *Evolution* 55(5): 849–858.

Ward, P. 2006. *Out of Thin Air: dinosaurs, birds, and Earth's ancient atmosphere.* Washington, DC: Joseph Henry Press.

Weinberg, S. 2000. *A Fish Caught in Time: the search for the Coelacanth.* New York: HarperCollins.

Wellman, C.H. 2004. "Dating the origin of land plants," in P.C.J. Donoghue and M.P. Smith (eds.) *Telling the Evolutionary Time: molecular clocks and the fossil record.* London: Taylor and Francis, pp. 119–141.

Whitehead, H. 2008. "Social and cultural evolution in the oceans: convergences and contrasts with terrestrial systems," in S. Conway Morris (ed.) *The Deep Structure of Biology: is convergence sufficiently ubiquitous to give a directional signal?* West Conshohocken, PA: Templeton Foundation Press.

Williams, G.C. 1966. *Adaptation and Natural Selection.* Princeton University Press.

Wimsatt, W. 1997. "Aggregativity: reductive heuristics for finding emergence," *Philosophy of Science* 64: S372–S384.

2007. *Re-engineering Philosophy for Limited Beings.* Cambridge, MA: Harvard University Press.

Wittgenstein, L. 1953/1973. *Philosophical Investigations*, Third Edition. London: Prentice Hall.

Woodward, S.R., et al. 1994. "DNA sequence from Cretaceous period bone fragments," *Science* 266: 1229.

Wray, G.A., Levinton, J.S., and Shapiro, L.H. 1996. "Molecular evidence for deep Precambrian divergences among metazoan phyla," *Science* 274: 568–573.

Index

Printed in the United States
By Bookmasters